"This book makes a critical contribution to environmental communication in India, a field of study much neglected, by offering a fine-grained analysis of communication strategies used by grassroots organisations in rural water management practices. Located within the framework of sustainable development and communication, the authors provide valuable insights into the forging of an incipient green public sphere in Bundelkhand and underline the role of women in the process."

– **Vinod Pavarala**, *Senior Professor, Department of Communication & UNESCO Chair on Community Media, University of Hyderabad, India*

"This volume documents how a grassroots organization is intervening in a drought-ravaged region of India and offers an important framework for understanding and interpreting environmental crises more globally. This is an essential text for examining communication strategies around dwindling natural resources, like water."

– **Gina M. Masullo**, *Associate Professor, School of Journalism and Media & Associate Director, Center for Media Engagement, The University of Texas at Austin, Austin, TX, USA*

"The book brings grave issues of water crises faced worldwide to the academic discourse. The environmental communication and civil society perspectives that it offers in analysing the environmental problems, particularly water-related ones in remote areas of India, make the text contextual and relevant for communication studies and also for social sciences."

– **Prof. Brij Kishore Kuthiala**, *Chairperson, Haryana State Higher Education Council, Panchkula, India*

"This work by Ram Awtar Yadav and Kanchan K. Malik on viewing Environmental Communication and Water Management in India from the prism of Civil Society is a welcome addition to the available literature on the broader theme. Adopting a case-study-based ethnographic approach, the study makes an important contribution to understanding a very complex theme. Approaching the water management challenges from the perspective of the efforts of non-governmental agencies who have involved local communities is truly a critical contribution. A must-read for all those working in the domains of Communication, Water Management and Civil Society, the work has successfully attempted a seamless integration of ideas and concerns."

– **Sandeep Shastri**, *Vice Chancellor, Jagran Lakecity University, Bhopal, and Professor of Political Science*

Environmental Communication and Water Management in India

This book analyses the underlying communication strategies and approaches of grassroots water management practices in India through a case study-based ethnographic approach.

Drawing from fieldwork experiences, this volume provides a detailed overview of Parmarth, a not-for-profit NGO, which is the case study for this research. It presents an in-depth theoretically informed analysis of data collected through multiple methods, which include key informant interviews, focus group discussions, participant observation, and document reviews, among other approaches.

The book examines Parmarth's strategies and processes to mobilise women as important stakeholders in the region's water conservation initiatives. It discusses communicative actions, tactics, and campaigns in water interventions and the role of various stakeholders ranging from local community members to civil society.

Accessibly written, this volume is a must-read for scholars and researchers of media and communication studies, environmental communication, ecology studies, development studies, public policy, sustainable development, water management, sociology, and political science.

Ram Awtar Yadav is Assistant Professor at the National Institute of Fashion Technology (NIFT), Raebareli, India. Before joining NIFT, he worked as an assistant professor with the Faculty of Journalism and Creative Studies at Jagran Lakecity University, Bhopal. He obtained Ph.D. from the Department of Communication, University of Hyderabad, India, in 2021. He did his Master's in Electronic Media from Makhanlal Chaturvedi National University of Journalism and Communication, Bhopal, in 2011, after which he worked as a journalist with *The Hitavada*, an English daily of Bhopal for around four years. He also worked as an assistant professor in the Department of Journalism and Mass Communication, Guru Ghasidas University, Bilaspur, Chhattisgarh, before embarking on his doctoral journey.

Kanchan K. Malik is Professor in the Department of Communication at the University of Hyderabad (UoH), India. She is Faculty Fellow with the UNESCO Chair on Community Media and Editor of the newsletter – *CR News*. Kanchan's Ph.D. from UoH contributed to policy advocacy by civil society groups for community radio in India. With a dual Master's in Economics and Mass Communication, she worked as a journalist with *The Economic Times* and as faculty in two reputable state universities before joining UoH in 2007. Her academic interests include Community Media and Women Empowerment, Journalism Studies, and Media Ethics. Kanchan was awarded the 2022–2023 *Fulbright-Nehru Academic and Professional Excellence Fellowship* to visit the University of Texas at Austin. She is a co-author (with Vinod Pavarala) of the much-cited book *Other Voices: The Struggle for Community Radio in India* (2007). Kanchan's recent co-edited volume (with Vinod Pavarala) is named *Community Radio in South Asia: Reclaiming the Airwaves* (Routledge, 2020). She published a manual last year titled *Strengthening Gender-Sensitive Practices and Programming in Community Radio* (UNESCO, 2021).

Towards Sustainable Futures

The Sustainable Development Goals (SDGs) were adopted by all United Nations Member States in 2015 as a universal call to action to end poverty, protect the planet, and ensure that all people enjoy peace and prosperity by 2030. The 17 SDGs are integrated, that is, they recognize that action in one area will affect outcomes in others, and that development must balance social, economic and environmental sustainability. The volumes in this series critically engage with the contemporary debates and discussions around development and sustainability. The series includes interdisciplinary and topical books presenting new perspectives for a sustainable future.

Sustainable Development Goals and Migration in South Asia
Edited by P. Sivakumar and S. Irudaya Rajan

Environment and Sustainable Development
Perspectives and Issues
Edited by Manish K. Verma

SDG 11, Sustainable Cities and Communities
Edited by Shayama Ramani and Hiroshan Hettiarachchi

Sustainable Development Goals and Indian Cities
Inclusion, Diversity and Citizen Rights
Edited by Ashok Kumar and D.S. Meshram

Future of Cities
Planning, Infrastructure and Development
Edited by Ashok Kumar and D.S. Meshram

Decent Work-Life in Business
Essential Tool for Sustainable Development
Soumi Majumder and Debashish Biswas

For more information about this series, please visit: www.routledge.com/Towards-Sustainable-Futures/book-series/TSF

Environmental Communication and Water Management in India

A Civil Society Perspective

Ram Awtar Yadav and
Kanchan K. Malik

Routledge
Taylor & Francis Group
LONDON AND NEW YORK

First published 2023
by Routledge
4 Park Square, Milton Park, Abingdon, Oxon OX14 4RN

and by Routledge
605 Third Avenue, New York, NY 10158

Routledge is an imprint of the Taylor & Francis Group, an informa business

British Library Cataloguing-in-Publication Data
A catalogue record for this book is available from the British Library

ISBN: 978-1-032-38579-2 (hbk)
ISBN: 978-1-032-39559-3 (pbk)
ISBN: 978-1-003-35028-6 (ebk)

DOI: 10.4324/9781003350286

Typeset in Sabon
Deanta Global Publishing Services, Chennai, India

To the water warriors of India and the world

Contents

Tables

Preface and Acknowledgements

This book is an outcome of our engagements with the issues related to water in various parts of India. With our close and strong affiliation to the field of communication, we analyse the underlying communication strategies and approaches of grassroots water management practices in the country through a case study-based ethnographic approach. This book particularly examines the water management interventions carried out by a local non-governmental organisation (NGO) of the Bundelkhand region named Parmarth Samaj Sevi Sansthan (PSSS) with the involvement and participation of the local communities living in the rural areas.

Through an in-depth analysis of its communicative acts, the book explores the working and approaches of Parmarth, its linkages with other organisations engaged in the water management sector, its networking with the governments, and its exchanges with the non-state actors. A distinctive aspect of this study has been the weaving of scholarly perspectives derived from theoretical frameworks of environmental communication, civil society, and communication for social change, with the narratives emerging from the lived experiences of the communities and civil society members, chronicled through extensive fieldwork carried out from the year 2017 to 2019.

The documentation and critical analysis of the insights from the field have been presented in this book through six comprehensive chapters. While the introductory chapter provides the context and background to the subject and case study, the second chapter documents the methodological terrain describing the research journey and actors involved in bringing this work together. The three thematic chapters narrate the story of grassroots water management based on analytical readings by authors of the voices of the stakeholders and participant observation notes jotted down meticulously during the study. The final chapter endeavours to argue for the imperatives of adopting a holistic and sustainable water communication approach to addressing grassroots water management issues.

We want to acknowledge all who have contributed to this study and its completion as a book. First and foremost, we owe a lot to the officials, staff, and volunteers of Parmarth Samaj Sevi Sansthan. We want to thank Mr. Sanjay Singh, Secretary, Parmarth, for all his support in providing access to

Parmarth's secretariat at Jhansi and other branch offices in the Bundelkhand region. Our gratitude also to the then state coordinators of Parmarth – Rupak Ghosh (Madhya Pradesh) and Manish Kumar (Uttar Pradesh) for their kind support during the fieldwork.

We also thank Parmarth staff members Manvendra Singh, Shivani, Ajay Yadav, Rajesh, Vineeta Yadav, Satyam, Sandhya, Abhinav, Deepal, Prashansha, Shahid Ali, Kailash ji, and others for facilitating the ground-level research in various ways. They were kind and generous in providing access to the required documents and arranging visits to specific intervention sites and events. It may not have been possible to interact closely with many of the beneficiaries of projects such as *Jal Sahelis* and meet activists and government officials associated with the initiatives without their help.

Several of the volunteers of Parmarth (like Uttam Yadav, Mihirlal Pandit, Dhaniram, and Naresh) served as a bridge with the community members in innumerable villages of Bundelkhand. We are extremely grateful to all the respondents who gave their valuable time (sometimes for hours) during the research, answering all the questions with immense patience and kindness. We salute their generous and welcoming attitude and all the formal/informal conversations during the fieldwork. We are also grateful to Pushpendra Bhai and Prem Singh ji from Banda for sparing their valuable time and discussing their work and the issues related to water, farming, and livelihood in Bundelkhand. Their inputs enriched the analysis in more ways than one.

We are also thankful to the leaders of the Jal Jan Jodo Abhiyan (JJJA) like Shri Rajendra Singh "Bhai ji," Shri Vinod Bodhankar ji, Shri Narendra Chug ji, Shri Maulik Sisodiyaji, Shri Jagdeesh Choudhary ji, and Major (Dr.) Himanshu ji, for being extremely kind and supportive and sparing time from their busy schedules for conversations around water issues. Some journalist friends in Madhya Pradesh also need special mention here who helped us identify and connect with the individuals and organisations working in the water management sector in Madhya Pradesh and Uttar Pradesh. Our sincere gratitude to Mr. Neeraj Santoshi Khar (Hindustan Times), Mr. Amarjeet Singh (The Times of India), Mr. Yogendra Singh Chandel (Dainik Bhaskar), and Mr. Sandeep Pauranik (IANS).

We feel indebted for the administrative support offered by the University of Hyderabad (UoH) and its Department of Communication (DoC) in facilitating the research that has now taken the shape of a book. Both DoC and UoH (an Institution of Eminence) not only offer a conducive atmosphere for independent and advanced research but also encourage scholars and faculty alike to present their work in various fora as well as benefit from the immense interdisciplinary academic resources available on the campus.

For Ram: All the teachers at the Department of Communication, University of Hyderabad, including the co-author of this book, Prof. Kanchan K. Malik, for being an inspirational collaborator, the Late Dr. P. Kennedy for all his insights on different occasions, and Dr. Arvind Susarla from the Centre for Regional Studies for his valuable feedback and comments during umpteen discussions on this work. My mentors and colleagues at Jagran Lakecity University, Makhanlal Chaturvedi National University of Journalism and Communication, Bhopal, and National Institute of Fashion Technology, Raebareli, for all the encouragement and support. My parents and siblings for always backing me in such endeavours, and my wife, Simpi, for all the unconditional support, love, and care and for happily bearing my prolonged busyness with such academic projects.

For Kanchan: To Ram, for working collaboratively throughout and keeping me engaged in all the nuances and interpretations of this research. I thank the team at UNESCO Chair on Community Media, University of Hyderabad, for the encouragement and ambience that it provides for good quality academic output. To friends and family – my daughter, mother, and sister – they make work and life meaningful.

Ram Awtar Yadav
Kanchan K. Malik

1 Unearthing the Communication Ecology of Water Management

Water, especially freshwater, is an essential element for humans as well as all other kinds of living beings on the earth. It is important for all forms of life on, in, and above the earth for their survival. According to the Hindu religious philosophy, the human body is made up of five elements – water (*jal*), earth (*prithvi*), fire (*agni*), space (*akaash*), and wind (*vaayu*). If we go by the modern sciences, water makes up about 60 to 70 per cent of the human body. That makes water one of the most important natural resources which affect our lives and health directly. It is the most basic need for human survival after the air.

History tells us that human civilisations have evolved around water bodies like rivers and lakes. The rivers and streams passing through human settlements provide various kinds of ecosystems and services to the people inhabiting them. They not only supply water to the people but also serve as systems that control floods and provide a balance to the natural environment. Water bodies, through civilisations, have been an integral part of the everyday lives of humans, and the rivers have nourished farming and agriculture for ages. Rivers have been the main channels for commerce and cultural exchange for centuries in human history. They have been considered holy entities and continue to be revered and worshipped.

In the last two to three decades, freshwater rivers have become carriers of garbage and sewage (*Down To Earth*, 2015). As revealed by the Central Pollution Control Board (CPCB) in March 2019,[1] at present, the water of the river Ganga, considered as the holiest river of the country, is not even fit for bathing, let alone drinking. Besides this, owing to issues like pollution, encroachment, damming, channelisation, etc., hundreds of small perennial rivers have either dried up or become seasonal. Similarly, innumerable lakes and ponds have disappeared, and many have become toxic due to high levels of pollution, causing a shortage of clean water for human societies.

Globally, around 1.1 billion people face water scarcity, while 2.7 billion population experiences a shortage of water at least once a month every year.[2] According to the Human Development Report 2019, the population of the world has, in the last 100 years, grown by around four times, while water extraction has almost septupled (UNDP, 2019). As a consequence

DOI: 10.4324/9781003350286-1

of the intensified human activities and climate change, India in the year 2019 faced its worst droughts in the past several decades (Sapru, 2019). According to a Government of India data, around 82 per cent of the country's rural households do not have piped water supply, whereas 163 million people do not have access to clean water in their close vicinity (NITI Aayog, 2019). The report also revealed that 70 per cent of the country's surface water is polluted.

As per the Global Climate Risk Index 2020, India is one of the most vulnerable nations in terms of the effects of climate change. Centre for Social Research (CSR), New Delhi, states that 90 per cent of climate-related events manifest through water in the form of drought, floods, hurricanes, etc.[3] India, indeed, faces the dual challenges of floods and droughts on a massive scale every year. While floods have become a common feature in various regions of India during every monsoon, droughts too are a recurring phenomenon during the summer season in different parts of the country. Around 600 million people face a high-to-extreme water crisis in the country every year (NITI Aayog, 2019).

Frequent droughts witnessed in several parts of the country have disturbing implications as manifested through challenges in sustaining agriculture and livelihoods in rural areas, farmers' suicides, large-scale migration, as well as environmental sustainability. The gap between the demand and supply of water is continuously widening. If we go by the scientific estimates, the water demand will surpass its supply by twofold by 2030, resulting in severe water crisis for millions of citizens (NITI Aayog, 2019). However, the good news is that looking at the seriousness of the situation, people have started working towards addressing the issues of water in numerous ways, and the interest of the societies in understanding water-related problems and finding their solutions is growing. This is happening at all levels – individually, collectively, and institutionally.

Today, India is witnessing many water conservation endeavours by individuals, communities, and civil society organisations that are aimed at addressing the issues of water crisis, and many of them have resulted in positive outcomes. Individuals and communities have transformed several water-scarce villages into water-sufficient ones in many parts of the country. The Bundelkhand region of India spread over the states of Madhya Pradesh and Uttar Pradesh is one of the highly drought-prone sites of the country that has experienced frequent droughts year after year in the last 20 years. In the last few years, lakhs of people have migrated to different places, abandoning their houses, villages, and farmlands due to the acute shortage of water that makes it extremely tough to sustain livelihoods. However, in many areas of the region, with efforts from civil society groups, communities have come together and have been working towards addressing the water crisis. As a result, for the last few years, thousands of water conservation structures have been created that have resulted in the availability of water for irrigation, drinking, and other household chores.

Although a significant amount of work is happening in the country in terms of environmental conservation in general and water management in particular that necessitates meticulous planning, coordination, communication, mobilisation, and resources, one rarely finds any substantial academic research in the area of environmental communication in the Indian context. This book, therefore, studies the communication strategies and practices as seen in water management initiatives being led by grassroots civil society organisations. In the process of carrying out this study, the book also explores the imperatives and discourses behind the grassroots environmental struggles around natural resources. It examines the nature of civil society interventions in water management to understand their interface with the state as well as local communities among which they work. It also studies the concept and approaches towards social change prevalent among the non-governmental organisations (NGOs) working in the water management sector and their communication for development approach, especially the theory of change adopted by them.

In the section following the statement of the problem above, the chapter presents an overview of the water-related discourses in the global context in general and India in particular. The third section highlights the workings of the water management sector in India and explains some significant interventions. This section elaborates on the water management interventions in the Bundelkhand region of India, which is also the site of this study. The fourth section provides a brief overview of the geographical location of the fieldwork and the details of the case study taken up for this research. The penultimate section of the chapter sheds light on the theoretical underpinnings that inform this study and the conceptual framework that has assisted in developing the analytical approach to this research. The final section of the introductory chapter presents a summary of the chapters that form a part of this book.

Water Discourses in India: An Overview

Currently, 40 per cent of the world's population is affected by water scarcity (Guppy & Anderson, 2017; World Bank, 2017). The global per capita availability of freshwater has declined by 55 per cent since the 1960s (Kashyap, 2019). Going by the current pattern of water consumption by various sectors, the water demand is expected to rise by 50 per cent by the year 2030 (UNESCO, 2012). By the year 2050, 2.3 billion more people are expected to be facing severe water crises mainly in the regions like North and South Africa and South and Central Asia (OECD, 2012).

As already discussed in the first section, the water crisis is triggered by two kinds of phenomena. First, the growing demand for freshwater for various human activities, and second, the depleting freshwater resources due to over-exploitation. The extraction of water from nature has increased at almost twice the rate of the growth in population in the last 100 years. As a result, many arid and semi-arid regions in the world are becoming water-scarce.[4] The fast-growing urban spaces are placing tremendous pressure on the water bodies around them, resulting in the aggravation of water scarcity. As the situation of water crisis worsens, governments across the world have started realising the impact it could have on human society in the days to come.

According to the *Global Risks Report* released by World Economic Forum for 2020, the water crisis is the number five global risk in terms of the impact that it will have on human society in the next ten years.[5] Going by the World Economic Forum's website, it is for the first time that the global risks report is dominated by the risks affiliated to the environment. The others among the top five risks are climate action failure, biodiversity loss, and extreme weather events. Barring the threat of weapons of mass destruction, which has been kept at number two among the risks to the world in terms of impact,[6] all others in the top five are environmental.

Considering the seriousness of water crisis looming large across the globe, several dedicated global organisations have been established to address the issue of water scarcity. Some such prominent organisations are UN-Water, the Stockholm International Water Institute, Global Water Leaders, the International Water Management Institute in Sri Lanka, World Water Council, World Resources Institute, World Wildlife Fund, Water.org, and others. There are various kinds of water management interventions that are being undertaken globally. Countries such as Israel, Saudi Arabia, UAE, and others have invested immensely in desalination of ocean/seawater, for meeting their water-related needs and are exporting desalinated water to other countries as well. Israel has also championed the process of recycling wastewater, and today it is recycling over 85 per cent of its wastewater. San Francisco in the USA has introduced laws to implement water conservation practices and made it mandatory for property owners and residents to use low-flow showerheads, water-saving faucets, toilets, etc. Cities like Boston and San Francisco in the USA also have energy codes in place which the large buildings have to abide by for reducing the consumption of water and energy.[7] They also have programmes to monetarily reward property owners investing in water and energy-saving systems.[8] Efforts are being made worldwide, both by governments and by civil society organisations, to safeguard and maintain the water conservation structures like lakes, ponds, wells, reservoirs, and other traditional systems suitable to specific geographical landscapes. Technologies for water metering are being used, like in the UK, to enable residents to understand the amount of water consumption for different uses and adjust it using online databases. Many countries are also using technologies for leak detection, adjusting water pressures, etc. for saving water. Rainwater harvesting systems are being installed in buildings in many parts of the world for recharging the groundwater by allowing rainwater to penetrate into the soil. Besides, many countries are switching to water-efficient agricultural practices to conserve water and using irrigation techniques like drip irrigation, sprinklers, efficient drainage, mulching (use of protective covers spread on the ground to reduce evaporation and maintain soil moisture), etc.

In addition to these, thousands of dams have been removed in the USA and Europe as part of some major initiatives for the revival of dying rivers. According to a report by *Down To Earth,* a total of 3450 dams have been demolished in the countries like Spain, the UK, Switzerland, France, Portugal, and Sweden, while the USA has dismantled around 1200 dams in recent decades to restore several rivers by ensuring their free and natural flow without any barriers.[9] Environmentalists all over the world are praising these moves by the governments of these countries terming them as "milestone" in ecological protection of the rivers and for bringing them back to life.[10]

In recent times, the environmental discourses, in India too, have been mostly dominated by the issues around water. The country has seen many protests against the damming of the rivers, and environmental activists in the country believe that the Indian government should take a cue from these countries and work towards protecting the Indian rivers by removing the dams on them and halting the proposed dams on the rivers. The Stockholm Water Prize Winner, and recipient of the Magsaysay Award, the well-known water conservationist, Rajendra Singh, who also served as a member of the Ganga River Basin Authority, says that hundreds of dams built on the river Ganga for irrigation and hydroelectric power generation are choking the river. Salopek (2019) too echoes similar views saying that over 300 big and small dams combined with the diversion weirs have led to diminishing flow of the river for the past few years. Many environmentalists also blame over-extraction of water, including groundwater, from the river's basin, as well as climate change, for the ebbing of the holiest river of the country. Today, half of India's rivers' water is considered unfit for drinking due to pollution caused mostly by the disposal of sewage and industrial waste in them. Toxic wastes, dead bodies of humans as well as animals, tonnes of cremation ashes, arsenic, and pesticides are slowly killing many of the Indian rivers like Ganga, Mahanadi, Cauvery, Yamuna, and others.[11] Like these rivers, several other resources for freshwater – lakes, ponds, tanks, wells, streams, etc. – have been dying for years.

Amid all these stories of rivers and other freshwater resources facing the threat of drying up, on account of the numerous human activities, there are many small but significant efforts being made by civil society groups, individuals, communities, and governments, among others, to restore the natural flow of rivers in the country. India, of late, has seen many resistances and initiatives by civil society organisations, local communities, and individuals in its different regions for the protection of natural ecosystems and the water bodies in it.

India has seen the emergence of several such movements and struggles that protest against moves that threaten the health of the rivers and the ecology around them. In the eastern region of the country there is the Save Dibang Valley movement in Arunachal Pradesh against a hydroelectric project of around 3000 megawatts proposed over the Dibang River that threatens the river as well as the 5000-hectare rich forest where over 2.5 lakh trees will be cut. Similarly, Manipur and Sikkim have seen various movements by local indigenous communities, civil society organisations, and activists demanding the new dams proposed by the government for hydropower projects to be stopped. Not only do these dams pose threats to the ecologically sensitive river valleys and their flora and fauna, but it will also lead to the displacement of local tribal communities in large numbers.[12] In the western part of the country, Nasik, in the state of Maharashtra, witnessed protests and demonstrations by several citizens' groups and NGOs, demanding de-concretisation of the Godavari River as the concrete embankments prepared during the Kumbh of 2003 were choking the riverbed and its banks. The activists took the issue to the court, and following the court orders, the Municipal Corporation of Nasik had to remove the concrete structures.[13]

In the north, several movements were and are carried out against the dams and tunnels for hydroelectric projects on the rivers like Satluj in Himachal Pradesh and Saryu, Kali, Ganga, Yamuna, Bhagirathi, Alaknanda, Ramganga, and others in Uttarakhand as they have damaged the ecologies of these river basins to a great extent.[14] In the southern part of the country, Kerala witnessed a widespread protest against the state government's decision to revive an unviable dam on the Chalakudy River. The decision faced opposition from the public as well as political parties, as a result of which, the Kerala Government had to give up the idea of this destructive project.[15] Similarly, the central region of India has long been known for the Narmada Bachao Andolan (NBA) (Save Narmada Campaign) led by Medha Patkar. In April 2020, Narmada Bachao Andolan (NBA) had a major victory as the Maheshwar Dam project on the Narmada River was terminated by the Government of Madhya Pradesh calling the project unviable and not in the interest of the public.[16] Similarly, another state of the central region of India, Chhattisgarh witnessed a strong movement by the tribal communities against a dam project on Indravati River after the central government gave the consent to take the work of the dam forward in May 2020.[17]

In addition to such protests and movements, the country has also seen several small but impactful and lasting initiatives for protecting water resources and creating new resources to conserve water.

Water Management Initiatives

The macro-level water management work has historically been the responsibility of the government. However, in the last two to three decades, there has been an emphasis on promoting people's participation in water resource management in the country. This section explains some of the recognisable water management endeavours undertaken by individuals,

NGOs, and other civil society groups in India. These include the initiatives spearheaded by (a) the Gandhian social activist, Anna Hazare in his village Ralegan Siddhi; (b) Popatrao Pawar in Hiware Bazar, Maharashtra; (c) Tarun Bharat Sangh in Rajasthan; and (d) Sant Balbir Singh Seechewal in Punjab.

Anna Hazare, by mobilising the community members of his village, especially the youths, did effective rainwater harvesting. He formed a group of youth in his village with a view to utilise the positive energies of the young population in water conservation work and named it 'Tarun Mandali' (group of young people). Besides this, he also mobilised women as well as people from the lower castes and carried out work to prevent rainwater run-off through watershed development, using local resources, especially voluntary labour by these people which ultimately reduced the cost of the interventions. The structures like contour trenches, farm ponds, check-dams, and afforestation carried out by the local communities under the guidance of Hazare transformed the Ralegan Siddhi into a model village in terms of water management practices.

Inspired by Anna Hazare, a Sarpanch (locally elected public representative), Popatrao Baguji Pawar, of Hiware Bazar village in the Ahmednagar district of Maharashtra, through effective rainwater harvesting and watershed development work transformed his impoverished village into a model village in around 20 years. Pawar, along with the villagers, also planted over one lakh trees around his village to turn it into a green village. His efforts were recognised nationally, and he received Padma Shri – the fourth highest civilian award in the country in the year 2020. Rajasthan-based NGO, Tarun Bharat Sangh, founded by 'Waterman of India,' Rajendra Singh, is a popular example of how NGOs are playing an important role in the field of water management in India. Led by Singh, the NGO, with support from the local communities of Rajasthan, has revived at least seven rivers of the state by constructing hundreds of check-dams and through extensive afforestation work. It has also transformed several villages like Gopalpura, Kishori, Hamirpur, Kraska, Bhaonta, Koylala, and many others by constructing *Johads* – an earthen traditional rainwater harvesting system in Rajasthan. The water management interventions in Rajasthan by Rajendra Singh were recognised globally and bagged him prestigious international awards like Magsaysay and Stockholm Water Prize, among others.

Similarly, the Sikh saint, Balbir Singh Seechewal of Punjab's Sultanpur Lodhi city restored the Kali-Bein River of the state which had turned into a sewage carrier and was on the verge of dying. Seechewal, who is also popularly known as 'Eco Baba,' started by educating the local communities about the importance of the river and gathered hundreds of volunteers who came forward to clean the river. Beginning in the year 2000, and after cleaning this river in a few years, the saint and his *sevadars* (volunteers) have taken on to clean and restore the other rivers in Punjab and are at present working towards de-silting the Satluj River.

The Dewas district of Madhya Pradesh was reeling under acute water crisis until the early 2000s resulting in a huge agrarian crisis. The district was supplied with drinking water via trains from the Indore city, and as such there was no question of availability of water for irrigation. In 2006, Mr. Umakant Umrao, the then Collector of Dewas, took stock of the situation by visiting hundreds of villages. He soon realised that the groundwater table of the region had dropped drastically and needed urgent revival. He interacted with hundreds of farmers in the district and convinced many of them to convert a part of their agricultural land into a pond for storing rainwater for irrigation. The farmers who adopted the idea started to reap the benefits immediately after the first monsoon. In around 4–5 years, over 10,000 ponds were developed by the farmers, and the water crisis issue was solved to a great extent bringing prosperity to the lives of the local people, especially the farming communities. The campaign led by Umrao was popularly known as the Dewas model of water management across the country and got global recognition too with the United Nations, in the year 2012, selecting it as one of the three best practices of water conservation in the world. There are countless such stories from different corners of the country wherein individuals and organisations have done significant works in terms of water management.

In many cases, the governments, both central and state, have carried out some successful programmes for water conservation and management. Notable among the initiatives by the governments, especially for improving the condition of irrigation, is Mission Kakatiya carried out by the Telangana Government. Mission Kakatiya worked for the restoration of the hundreds of existing minor irrigation structures like ponds and tanks with the support of the local farmers and Gram Sabhas (village councils) that enabled thousands of farmers in the Godavari and Krishna river basins to have water for irrigating their farmlands. Similarly, the Narmada Canal Project by the state government in the Jalore district of Rajasthan that made use of sprinklers and made drip irrigation mandatory transformed the farming scenario in the area and benefitted the local farmers immensely. Likewise, the implementation of *Har Khet Ko Pani* (water to every farm) scheme run by the Ministry of Water Resources, River Development and Ganga Rejuvenation (now, Ministry of Jal Shakti) in the Chittoor district of Andhra Pradesh resulted in the restoration of a chain of tanks with support from community members. This initiative saw the participation of around 1400 community-based organisations (CBOs). It became a model and was recognised by the Government of India in the year 2017 as among the best practices in irrigation systems in the country. There are stories of several such successful government interventions from other states like Gujarat, Kerala, Odisha, Maharashtra, and others.

Several noticeable initiatives related to drinking water management have been undertaken by NGOs, individuals, and civil society groups along with government agencies in the states like Uttarakhand, Uttar Pradesh, Madhya

Pradesh, Gujarat, Rajasthan, Kerala, and others. This book is a critical study of one such water management intervention carried out in the Bundelkhand region of Uttar Pradesh and Madhya Pradesh.

The Site of the Study

The Bundelkhand region of Central India is characterised as semi-arid, receiving very little annual rainfall, because of which it has become one of the hotspots of water crisis in the sub-continent. Several studies reveal that the region has been experiencing a declining trend in annual rainfall for the past few decades (Ahmed et al., 2019; SANDRP, 2018). In the last two decades especially, the region, due to a continuous deficit in yearly rain, has had to cope with back-to-back droughts resulting in acute shortage of water for agricultural purposes as well as drinking and household usage.

Looking at the harsh living conditions in the region triggered by water scarcity, land degradation, large-scale migration, and poor socio-economic conditions, several development organisations, NGOs, and local groups have come forward to work towards addressing the issue. Different organisations and individuals are undertaking various kinds of initiatives in several pockets of the region. Many international organisations are also carrying out projects related to water conservation in the area in association with local NGOs and government agencies.

This study analyses the water management interventions undertaken by Parmarth Samaj Sevi Sansthan (hereafter mentioned as Parmarth), a local NGO with its headquarter in the Orai town of the Jalaun district of Uttar Pradesh, secretariat at Jhansi, and branch offices in different areas of the region both in Uttar Pradesh and in Madhya Pradesh. The book also delves into the activities and events being organised by Parmarth as a part of its nationwide campaign called *Jal Jan Jodo Abhiyan* (JJJA), meaning water people alliance campaign. The JJJA campaign by the NGO aims at bringing the various stakeholders together, viz. civil society organisations, government, activists, academics, policymakers, and politicians, to push for a law ensuring the right to water and protecting the rivers and other water resources in the country from dying. This research studies the water management initiatives by the NGO in two districts of Bundelkhand region coming in Uttar Pradesh (UP), i.e. Jhansi and Lalitpur and two districts of Madhya Pradesh (MP), namely Chhatarpur and Tikamgarh.

Theoretical Background

In the past few decades, environmental concerns have taken the centre stage in major global discourses. The fundamental elements that humans and other living organisms need like clean air and water are getting scarce by the day. Tonnes of toxic chemicals are being produced every day, polluting the air and water resources. Lakhs of trees that provide air to breathe are being cut every day.[18] And the earth is warming up predominantly due to anthropogenic activities. The climate is changing, causing altered weather patterns in different regions of the world that are also encountering uncertain weather conditions. The frequency of extreme weather events has increased manifold in the last few decades. Water resources are getting depleted, and some are disappearing. With all these environmental crises surfacing before the world in contemporary times, societies are seriously engaging with the issues related to environment, and efforts are being made at all levels by various stakeholders to find effective solutions. Based on these premises of the need for engagements and multi-stakeholder partnerships to address environmental issues, this text draws from three major theoretical frameworks to study civil society-led water management practices in India. These are environmental communication, communication for development, and the concept of civil society.

Environmental Communication

Environmental communication is considered a discipline of crisis that is located in the global context of the ever-rising need for taking measures to contain the environmental problems facing the world today (Cox, 2007). As a definable field of academic enquiry, it emerged within the broad field of communication in the last three decades (Cox & Depoe, 2015) coinciding with the period when environmental degradation and related concerns shot to the peak worldwide. Cox and Depoe (2015) define environmental communication as a field of study that "strengthens the capacity of societies to deliberate and respond to conditions relevant to the well-being of both society and natural biological communities" (p. 15). Early scholarship in the field of environmental communication was mostly about the representation of environmental issues by the media and analysis of the rhetorical texts. In contrast, recent works in the field have widened the domain scope that ranges from discourses on sustainability, climate change communication, food, water, the binary between culture and nature, environmental justice, agriculture, and others (Retzinger, 2010; Sowards, 2012).

For Pezzullo and Cox (2018) when defined appropriately, environmental communication as an academic discipline takes into account a wide variety of areas for studying that include "language, visual images, protest, music, or even scientific reports as different forms of symbolic action" (p. 34). The authors draw from Kenneth Burke (1966), who says that both our languages and other symbolic acts communicate something to others. In this particular study too, the authors find that the languages, as well as the other symbolic acts as mentioned here including images, protests, music, and scientific studies, play out in some way or the other for sharing information or talking about the environmental issues (mostly related to water in this case).

One of the most important characteristics of the field of environmental communication in the recent past has been its focus on the enhanced level of public participation, especially by ordinary people, activists, environmental groups, civil society organisations, businesses, and others in the significant decision by the governments that affect their lives and future (Pezzullo & Cox, 2018). And for any participation to take shape, the element of communication is fundamental (Singhal, 2001). The ongoing and looming water crisis in the world has become one of the most important topical foci within environmental communication, and issues like climate change and global warming. When the world is facing a severe ecological crisis that threatens the future of humanity, Ahmad (1992) feels that the need for people's participation in managing natural resources is paramount. It is based on such theoretical understandings that this book attempts to analyse the water management initiatives of Parmarth in the Bundelkhand region of India where the NGO, with active community participation, has, to a great extent, addressed the water-related issues in its intervention areas. The book investigates how, through the communicative actions of Parmarth, like training programmes and meetings, the communities gain significant insights into their role as actors of the change-making efforts. They go on to engage in dialogues with the local government authorities to advocate for water issues that affect them and seek their interventions in addressing those problems.

Communication for Development

Since the term 'development communication' became accepted in the 1970s, it has been defined in different ways by different scholars, all referring to the socio-economic transformation of societies through the use of mass media while ensuring sustainability. Earlier, development communication was primarily understood as communication occurring through mass media to educate people to attain growth and development through their partnership. In his book, *Mass Media and National Development*, in 1964, Wilbur Schramm, whose work is among the milestones in the history of development communication, underlined the importance of communication in the process of socio-economic development and said, "Without adequate and effective communication, economic and social development will inevitably be retarded, and may be counter-productive" (Schramm, 1964; p. 6).

As the discipline evolved, the western models of development were criticised by many scholars who felt that socio-economic and cultural diversities in Latin America, Africa, and Asia demanded development models which were in line with their local needs. They felt that the replication of the western development models would not redress the issues at the grassroots levels in these regions of the world. As such, the non-western countries came together as Non-Aligned Movement (NAM) wherein they tried to be self-dependent and come out of the clutches of the developed western nations (Servaes & Malikhao, 2002). However, these models were still focused on the dominant top-down modernisation and knowledge dissemination approach aimed solely at poverty alleviation.

In the 1970s, the dominant paradigm of development communication was further subjected to critique from various scholars who believed that communication for development (C4D) models should be aimed at attaining equity and social justice in addition to economic growth (Baba, 2015; Moemeka, 1994). From the late 1980s onwards, the notion of development communication began to change. The focus started shifting from one-way mass media–oriented development communication to the one which aimed at the participation of all stakeholders stressing the importance of holistic and sustainable development of local communities brought about through democratisation and participation of social change processes at all levels – international, national, local, and individual (Servaes, 2008).

Later, social and behaviour change communication (SBCC) approach to communication for development emerged, which emphasised that the nature of communication "should be empowering and horizontal" and should:

> encourage communities to be agents of their own change; promote dialogue, debate, and negotiation (as compared to information and persuasion techniques); emphasise the process of interactions, shared knowledge, and collective action (rather than a sender-receiver model); and focus – beyond but to include individual behaviours – on social norm change, policies, and culture to unfold sustainable change in communities and among individuals.
>
> (McKee et al., 2014; p. 278)

The civil society organisations, NGOs, and other groups working at the grassroots level to attain such goals through the SBCC approach of C4D also built their own theory of change based on important questions like, what they want to do and how (Care Nepal, 2012). The theory of change, which of late has become an important dimension of C4D, provides "a very powerful learning lens" through which the NGOs can "develop a clear framework for monitoring and evaluation; more common understanding, clarity and effectiveness in their approach; and strengthen their partnerships, organisation development and communication" (James, 2011; p. 3). The significance of the theory of change by the development practitioners and organisations is also underlined by Patricia Rogers (2014) as she calls it "a building block for impact evaluation" (p. 3) after which certain practices could be applied to another site where the change is sought. This study also draws from Tufte (2017) who stresses the need to embed and review communicative efforts of the organisations in the development and social sector from a broader socio-cultural perspective of subject and institution instead of just analysing their communication strategies. For a deeper understanding of the relationship between communication processes and social change, researchers must tread the often neglected path and explore communication practices of the social actors as a social, relational, and dynamic process (Tufte, 2017). Such scholarship helped the authors delve deeper into the dialogic communication processes that occur face-to-face at the grassroots level between the development practitioners and the local communities. These processes lead to the capacity-building of the communities who learn to advocate for their rights before the local government and administration and improve their livelihoods.

Thus, this study strives to gauge the C4D approaches prevailing in select water management projects being implemented by Parmarth and its theory of change. Through the theoretical lens of C4D, predominantly its theory of change approach, this study examines the communication strategies and practices adopted by Parmarth for working with the communities to bring about behaviour change, especially in a gender-responsive manner. It also attempts to figure out the theories of change that come into play as Parmarth partners with the women in the villages to make them important stakeholders in its water-related interventions. Many times, as described by Manyozo (2012), the lack of mutual exchange and reliable communication networks among rural communities prevents them from representing themselves in the society. Such a lack of exchange and communication networks could be gauged in the Bundelkhand region. People belonging to different castes and occupations could be seen maintaining a social distance from each other. In the region, one could observe that there are wells and hand pumps designated to different castes in many villages. In some villages where there is one well to draw water from, the local communities use different bearing pulleys for different castes for extracting water. Against this backdrop, the research examines how grassroots organisations like Parmarth have played a significant role in mitigating caste differences and facilitated the formation of reliable communication networks by bringing people from different castes together and representing their issues before the state and its agencies. This helps them get their issues addressed and empowers the marginalised communities belonging to different castes to forge alliances and to seek solutions to their local problems, especially water-related woes, with cooperation and collective action. Such coming together of communities also helps the grassroots organisations engaged with development interventions seek their participation effectively which – as Robert Chambers writes in his foreword to Shirley A. White's (1999) book – makes projects economical as well as more effective. These communities, then, also take responsibility for the projects and make sure that they become a sustainable resource for them. Stressing upon the need for effective grassroots communication, Anyaegbunam et al. (1999) talk about a communication approach called Participatory Rural Communication Appraisal (PRCA). PRCA refers to a two-way transactional process that enables field workers and mobilisers to start a dialogue with community members in a given setting and involve them in processes that lead to mutual understanding and action. Anyaegbunam et al. (1999) believe that PRCA, unlike other conventional communication approaches, enables the mobilisers to create alternative ways of designing messages for the local communities who sometimes share different cultural norms, experiences, values, and traditions than the facilitators and discover strategies that empower the communities to create their own messages. Building upon these theoretical entry points, this book unearths strategies for facilitating grassroots communication and the stories of change through the collective endeavours of the facilitators and community members, especially women, in the rural areas of the Bundelkhand region.

Civil Society

To a great extent, the development sector has come into the hands of many civil society organisations and individuals from government-owned practices (Kamat, 2002). For Kamat (2002) the proliferation of the grassroots organisation indicates an emergence of a political space that is diverse from the state's notion of development as the latter neither ensures popular participation in a development intervention nor people's empowerment. However, Wood (1997) talks about the new kind of relationship that is developing between the state and the NGOs in the neoliberal context and calls the latter 'franchise state' wherein they take up the management and administration of the essential social services from the governments and implement them with their own resources and infrastructure. When it comes to effective water management in India, the contribution of the local communities and NGOs is considered extremely significant. Mihir Shah Committee on Restructuring of Central Water Commission (CWC) and Central Ground Water Board (CGWB) in its report (2016) emphasises that "Water resource management has to be a participatory process with all stakeholders (including NGOs) and local communities (particularly farmers at the Panchayat level) who should be made aware of the water resource issues and who need capacity building for management of the limited water resources (e.g., participatory irrigation management, appropriate crop selection, micro-irrigation, conjunctive use, wastewater reuse/recycling, etc.)" (p. 85). The committee also underscores the civil society organisations' role in such capacity-building activities, as they hold experience in the management of natural resources and sustainable development, especially when it comes to training the communities.

In the quest for a good society, collective action has always been a part of the universal human experience manifesting in innumerable ways (Edwards, 2004; Parekh, 2004). The concept of civil society, as Edwards (2004) feels, has occupied the centre stage of the development discourses in the last two decades as millions of people across the globe find in it an inspiration for their struggle in search of a better world. Drawing from the idea of civil society that facilitates the mediation between various actors and offers women and men a space to carry out voluntary and public activities (Barber, 1995), the book explores the role of Parmarth in uniting various stakeholders together to work around the issues of water through its national level campaign that aims at building a strong civil society around water. Similarly, taking a cue from scholars like Mahajan (1999) and Beteille (2001) that civil society and state actors should share a constructive relationship in a democratic system, this work examines the nature of interactions between the campaign leaders and governments both at the centre and in the states on the issues of water. Another important entry point for this study is the idea of Ahmad (2018) that a "kicking and thriving civil society help foster an atmosphere of transparency and accountability through the emphasis on public awareness and participation in the governance system" (p. 17). Concurring with this view, this research endeavours to study the efforts of the civil society in India to hold the governments accountable for their responsibility to ensure water security for all citizens by facilitating public awareness and participation on the issues of environment and water under the JJJA Campaign of Parmarth.

The book also draws from Giri (2012), who considers civil society a conscious and mobilised aspect of society which creates a space for critical self-reflection and public deliberation. A similar process and approach was observed in the various consultation camps and other deliberations organised at regular intervals by Parmarth under its JJJA campaign. By applying this understanding of the concept of civil society, the study analyses the various consultations and meetings organised by Parmarth to deliberate and discuss the issues of water scarcity in various parts of the country and how they could be addressed with support from the public as well as the governments.

Outline of the Book

This text is an ethnographic exploration of water management initiatives and the communication processes embedded in them in the rural hinterlands of the Bundelkhand region of India. While the interventions are carried out by an NGO called Parmarth, the commitment of the local communities and, in some cases, local government officials is also equally significant in successful water management projects. This section provides a comprehensive description of the structure of the book by presenting an overview of the chapters through a brief introduction of their content. The book consists of six chapters including this introduction, followed by a chapter on methodology, three core thematic chapters, and the conclusion.

After this introduction chapter which presents the background to the book and an overview of its structure, Chapter 2 discusses the methodology used for carrying out this research. It narrates the authors' journey into the field that began with the pilot study and how they selected the final case study for immersive fieldwork. The chapter then lists out the research objectives that this study set out to achieve and the research questions which the authors seek answers for through this work. It then outlines the research methodology, which is qualitative, and employs ethnographic case study design. It also discusses the methods used for data collection, including in-depth interviews, focus group discussions (FGDs), participant observation, document analysis, go-alongs, and the overall description of the fieldwork done. The chapter also discusses the process of analysis preceded by coding the data gathered through the tools and methods mentioned above. It then provides an overview of the Bundelkhand region of India where the study is based. The chapter includes the profile of Parmarth which is the case study for this research wherein the authors enumerate different kinds of interventions being carried out by the NGO in Bundelkhand and a campaign being run by it at a national level.

Chapter 3 studies the theory of change developed by Parmarth and its application in the interventions related to different projects. The chapter draws from scholarship on the theory of change and analyses the strategies and processes that go into mobilising women as important stakeholders in the initiatives related to water conservation in the region. The chapter explains how Parmarth starts most of its interventions by carrying out a baseline study through participatory rural appraisal (PRA) tools involving local communities. This exercise is carried out to understand the existing situation and map the severity of the existing issues and challenges. As a part of its initiatives to address water-related problems, Parmarth also makes efforts to capture the social norms and practices in the proposed intervention areas to make effective strategies to deal with the social challenges underlying a programme implementation and see that the desired changes take place by a change in social behaviour. The chapter then talks about how the organisation identifies the relevant stakeholders and how it mobilises them to act for seeing the desired change. It also discusses women as important stakeholders and the formation of the *Pani Panchayat* groups led by women in the villages. The next section of the chapter outlines how change happens through participation and engagement which ultimately leads to empowerment of women and other community members. In the final section, the chapter discusses how the *Jal Sahelis*, who are the leaders of the *Pani Panchayat* groups, sustain the change and take the work forward with their collective efforts and by ensuring support and participation of the community members as well as the other stakeholders.

In Chapter 4, the authors present the literature for understanding and defining the field of environmental communication and then describe how NGOs working on environmental issues adopt and practise environmental communication. The chapter then specifically considers the case of Parmarth and outlines its programmes and activities in terms of environmental communication. It examines different programmes run by the NGO like the watershed development programme in the Chhatarpur district of Madhya Pradesh and how the village-level committees are formed to ensure people's participation in the projects and in the important decisions for carrying out the related work. The chapter also explores the various kinds of activities carried out by Parmarth for mobilisation and communication about water among the local communities of Bundelkhand. The chapter also discusses how some of the women in the region take leadership roles after the training and capacity-building programmes and become great mobilisers and communicators for garnering communities and other stakeholders' support for water-related work. It also discusses the other projects of the NGO like the WADI project which is about helping the poor tribal community of the Talbehat block of the Lalitpur district of Uttar Pradesh develop fruit orchards and forestry for livelihood enhancement. The chapter also describes the kitchen garden practice promoted by Parmarth among the poor communities of the area for meeting their nutritional needs and making use of the used water.

Chapter 5 analyses, in detail, the JJJA Campaign of Parmarth and the activities being carried out across the country as a part of this campaign. The campaign aims to bring the society and government together for water conservation, protecting water bodies, and rejuvenating dying rivers through a combination of traditional and modern methods. The chapter closely examines how the JJJA campaign is working in bringing all the stakeholders from different parts of India together, facilitating dialogues among them, improving water literacy and awareness among the people, and most importantly, pushing for a law ensuring water security for all in the country on the lines of the food security law. Through this chapter, the book traces the genesis of the campaign that started with regional consultation meetings in different states like Madhya Pradesh, Bihar, Rajasthan, Karnataka, Kerala, and a few others. The campaign first identified the important stakeholders in these regions including the activists, water practitioners, social workers, scientists, academics, and intellectuals to initiate discussions among them about the issues of water crisis in India and their probable solutions.

In Chapter 6, the authors discuss the important aspects of the different chapters and pull them together to draw deeper insights that cut across the varying thematic areas in this research. This chapter provides an overview of the diverse approaches visible in the interventions of Parmarth, communicative actions, strategies, and practices that underlie these interventions and the role of various stakeholders ranging from the local community members to the civil society organisations working in the field of water. Through assimilation of the important insights and reflections drawn from this research, the conclusion brings forth the ideas around 'water communication' that demand special attention within the field of environmental communication. It argues that considering the way the issue of water shortages is magnifying across the globe, and the governments and civil society are engaging with these concerns, 'water communication' could emerge as one of the major thrust areas demanding an increased level of engagement from both academicians and practitioners within the communication discipline. Water security is considered as the key element for the attainment of sustainable development goals. The conclusion chapter, therefore, also brings to the fore some other important insights that emerge from this study like the need for a holistic communication approach for closer cooperation between different stakeholders required for the attainment of the sustainable development goals, and the resonance of Gandhian ways of communication as reflected in communicative actions by civil society organisations for their work related to environment and water.

Notes

1 https://tinyurl.com/y5n3mnss. Accessed on October 15, 2020.
2 https://tinyurl.com/yyttnlfe. Accessed on November 4, 2020.
3 https://www.csrindia.org/women-and-environment/. Accessed on November 5, 2020.
4 https://www.constrofacilitator.com/different-types-of-water-management-methods/. Accessed on November 8, 2020.
5 https://tinyurl.com/y4u5s2gf. Accessed on November 8, 2020.
6 Ibid.
7 https://www.water-io.com/top-cities-breaking-the-mold-in-water-conservation. accessed on November 9, 2020.
8 Ibid.
9 https://www.downtoearth.org.in/news/water/by-demolishing-dams-europe-us-letting-rivers-reclaim-their-space-60572. Accessed on November 10, 2020.
10 Ibid.
11 https://www.deccanherald.com/content/653974/toxic-wastes-dams-killing-our.html. Accessed on November 10, 2020.
12 https://sandrp.in/2020/09/26/world-rivers-day-2020-celebrating-rivers-across-the-world/#more-37970. Accessed on November 12, 2020.
13 Ibid.
14 https://tinyurl.com/ycdkws5l. Accessed on November 12, 2020.
15 https://sandrp.in/2020/09/26/world-rivers-day-2020-celebrating-rivers-across-the-world/#more-37970. Accessed on November 12, 2020.
16 https://tinyurl.com/yxodttcp. Accessed on November 12, 2020.
17 https://tinyurl.com/y9c7wzms. Accessed on November 12, 2020.
18 https://www.reference.com/science/many-trees-cut-down-day-42bf5e6262028f2d. Accessed on November 14, 2020.

2 Analysing Water Management from an Environmental Communication Perspective

A Qualitative Approach

This study is an endeavour to closely document and critically analyse the grassroots interventions in water management by an NGO in the Bundelkhand region of India, as well as a national-level water-centred campaign by them from a communication perspective. A significant amount of water conservation work has been done in several pockets of the Bundelkhand region of India, which is known for frequent droughts and subsequent migration. Following a pilot work carried out in February and March of 2017 in Madhya Pradesh and again in April of 2017 in the Alwar district of Rajasthan, the water-related interventions of Parmarth Samaj Sevi Sansthan (henceforth referred to as Parmarth) and its advocacy efforts with the State were chosen as a case study for this research.

As stated in the introduction, this study applies different theoretical lenses to study the community-based water management initiatives that have been undertaken in different parts of the Bundelkhand region of Uttar Pradesh and Madhya Pradesh supported by Parmarth. It also examines their campaign called "*Jal Jan Jodo Abhiyan*" (JJJA) which translates as "water-people alliance campaign," and is a nationwide movement spearheaded by Parmarth with support from other civil society organisations and activists from across the country. This campaign seeks to mobilise various stakeholders towards protecting the water bodies in the country by pressing for a law to ensure water security for all.

The continuously rising demands for water in every sphere of human existence, dying water bodies, and deteriorating quality of water are making water crisis a pervasive phenomenon globally. The supply of per capita potable water is shrinking fast. The Bundelkhand region of India, which was once known for its thick forests attracting dense rain-bearing clouds, has today become synonymous with recurring droughts and water crisis. The initiatives by Parmarth at local levels to address the issues of water scarcity by mobilising support from the local communities and governments, thus, provided the authors with a suitable case for immersive academic research. The research aimed to explore how Parmarth communicated about the problems to various stakeholders and sought solutions through their participation. Parmarth took up the task to build awareness among the local

DOI: 10.4324/9781003350286-2

populations in multiple areas of Bundelkhand about the causes of water crisis and how people could contribute their share of effort to deal with the situation. Simultaneously, it started bringing other civil society actors from various regions together on national level so that they could strongly advocate for policies for protecting big rivers and their tributaries throughout the country.

An extensive fieldwork was carried out to collect the relevant data related to the local water conservation initiatives undertaken by Parmarth in numerous villages of the Bundelkhand region. The authors also participated in several programmes organised by Parmarth under the *Jal Jan Jodo Abhiyan* (JJJA) at the regional and national levels and observed the processes employed by them to facilitate dialogues among various stakeholders from different sections of the society to come up with effective ways of addressing the cause of water crisis in India.

The following sections of the chapter first outline the research objectives and questions that guided the study, and then describe the research methodology adopted. There is a discussion of the case study method chosen for this research, and the chapter also elaborates on the data collection tools used for the case study like in-depth interviews, focus group discussions (FGDs), participant observation, and document analysis. Then, it goes on to describe the fieldwork and the coding and analysis process of the research. Importantly, towards the end, the chapter presents the profile of the Bundelkhand region of India and of the organisation being studied as a part of this research, i.e. Parmarth Samaj Sevi Sansthan (henceforth referred to as Parmarth).

In order to systematically prepare the roadmap for this study, research objectives and questions were developed following several rounds of discussions amongst both the authors.

About the Study

The objectives and questions listed below informed the process of identifying the potential cases for studying and finalising one in the end. These were also helpful in informing the data collection methods, approach, and analysis.

1. To study the imperatives and discourses behind grassroots environmental movements and struggles around natural resources in India and their linkages with the issues of development and identities.

   What are the contemporary discourses nurturing grassroots environmental movements in India?

   What are the issues related to development that propel struggles around natural resources?

   How are struggles around natural resources also associated with issues of culture and identities?

2. To examine the nature of civil society interventions for water management, especially in remote areas in the context of their interface with the state as well as the local population.

   What are the different types of civil society groups and organisations active in the water management sector, and what is their modus operandi?

   How do the organisations working in the water management sector network with the government at the national and state level?

   What are the factors that determine the geographical region and communities that these organisations work with?
3. To identify the concept of and approaches towards social change prevalent in the NGOs working in the water management sector and the resulting Communication for Development (C4D) practices adopted by them.

   What are the organisational structure, management systems, and decision-making practices adopted by the NGOs working in the water management sector?

   What is the philosophy behind the social change and mobilisation approaches adopted by the NGOs working in the water management sector?
4. To analyse the communication strategies and designs adopted by the NGOs working in the water management sector to interact and work with their stakeholders, such as the communities, government, bureaucracy, and corporate houses.

   What is the Theory of Change (ToC) constructed by the organisation for bringing about social and behavioural change and collective action at the grassroots level?

   What are the communication strategies used by the NGOs to mobilise and involve communities in the various stages visualised within its theory of change model?

   What are the levels of engagement among the community members in the areas where the NGOs are implementing water management projects?

   How do the NGOs seek the involvement of other stakeholders like the government, bureaucracy, and corporate houses in their water management initiatives?

   How do NGOs engage in activism and advocacy around issues related to water management while simultaneously depending on the government, bureaucracy, and corporate houses to implement and sustain their projects and initiatives?

Collecting the Data

This research employed qualitative methods of data collection for the fulfilment of the research objectives of the study. The design and nature of the research were such that it required more than one visit to the same places by the researchers to engage and interact with the stakeholders and observe the progress of the initiatives being undertaken by the NGO and the community members. Thus, they followed an ethnographic approach for gaining an in-depth understanding of the context and realities of the interventions around water and for obtaining insights from the respondents at different points in time. This research uses the case study method wherein the ways of data collection included in-depth interviews, focus group discussions, participant observation of the events and activities, and informal interactions and go-alongs with various relevant respondents in the field. Besides this, an analysis of several types of documents produced by the Parmarth and the JJJA campaign was also carried out for a deeper comprehension of their programmes.

The authors directly met and interacted with various stakeholders of the water management initiatives in Bundelkhand and travelled to different places within the region for developing an understanding of the existing situation in terms of availability of water and the people's daily struggles for fulfilling their needs of water where it is scarce. It was this process that led to the decision to take up Parmarth Samaj Sevi Sansthan (PSSS) as the case study and to closely examine its water management interventions such as the water-shed development projects in the Chhatarpur district of Madhya Pradesh, and the initiatives of reviving old water bodies like lakes, ponds, wells, etc. in various districts of Uttar Pradesh included in the Bundelkhand region. Efforts were made to ensure that this research captures the experiences of all kinds of stakeholders from the office-bearers of Parmarth to the leaders and representatives of village-level committees. It also sought to include the perceptions of groups constituted in various villages as well as the volunteers and other activists who contributed to taking the initiatives of water conservation forward.

The qualitative data for this research was gathered in the form of people's responses and perceptions as well as the observations of realities on the ground. These were recorded as thick descriptions and reflections in the notes during the fieldwork. Excerpts from the debates and discussions at various programmes and events organised by Parmarth in Bundelkhand on different occasions and under the *Jal Jan Jodo Abhiyan* in different parts of the country also formed a substantial amount of data. The insights drawn from the various kinds of documents related to the interventions and advocacy efforts were also recorded in the field notes for analysis to be carried out later.

Methodology

As Schensul (2008) puts it, "Methodology consists of the actions to be taken in the study and the reasons for these actions in testing or generating theory" (pp. 16–17). Research methodology could also be described as a roadmap for conducting a study and choosing the appropriate research methods which ultimately helps in selecting the right tools and techniques for collecting data and analysing them. In this research, a case study research design was used for the data collection. The case study method lends itself to diverse orientations as proposed by Hancock and Algozzine (2016), which include ethnographic, historical, psychological, or sociological. This study employs the ethnographic case study design, suitable for exploring the observable patterns of behaviour, everyday life experiences, and customs in a group sharing a culture.

Hancock and Algozzine (2016) explain, "Ethnographic case studies typically involve extended interaction with the group, during which the researcher is immersed in the day-to-day lives of group members" (p. 31). In this study, the researchers spent a significant amount of time interacting with members of communities around the sites where the water management initiatives were carried out and also with the volunteers and personnel of Parmarth. The researchers spent extended periods staying along with the workers of Parmarth in different parts of Bundelkhand like Chhatarpur, Jhansi, Tikamgarh, Lalitpur, Jalaun, and others. Staying in the field not only enabled the researchers to have prolonged interactions with the staff of Parmarth but also with the community members from the local villages.

According to Gobo (2008), ethnography consists of at least three terms, namely "participant observation," "fieldwork," and "case study" (p. 11). In this research, all these three terms form integral parts of the research methodology. While "case study" is the research method applied in this study, "fieldwork" features as a data collection process and "participant observation" is employed as one of the data collection tools. All these terms are discussed in detail in the following section of the chapter.

Case Study Method

The case study method is a preferred research strategy when a researcher is attempting to explore and understand some processes in a real-life context (Yin, 2003). It is considered useful in carrying out an intensive enquiry of a particular case (Gerring, 2007) wherein through detailed observation with a combination of tools like interviews and document analysis a study could explain how specific processes happen in a given setting (Verschuren & Doorewaard, 2010). Drawing from Yin (2003; p. 1) that a case study could be helpful for research, especially when attempting to understand the "individual, group, organisational, social, political, and related phenomena," this research applies this method to study a Bundelkhand-based NGO for its water-related interventions. As mentioned earlier, these interventions are based in numerous villages of the region that face a dearth of water due to various reasons ranging from an errant monsoon to over-exploitation of groundwater resources and poor maintenance of the traditional water conservation structures.

According to Hancock and Algozzine (2016), the subjects of enquiry under a case study may vary extensively, and a researcher may attempt to gain insights from an event, situation, programme, or activities applying this research method. Data for carrying out a case study could be gathered from various sources which may include documents, interviews, ethnography, surveys, audio-visual materials, field notes, observations, archives, physical artefacts, etc. (Gerring, 2007; Hancock & Algozzine, 2016; Yin, 2003).

This research too uses some of the methods mentioned above for collecting evidence through documents, participant observation of the events and activities, direct observation, field notes, in-depth interviews, focus group discussions (FGDs), and others.

In-depth Interviews

Interviews are a common method of collecting data while doing a case study research, and when conducted face-to-face, it could bring richness to the communication between the interviewer and the respondents (Gillham, 2000). In-depth interviews enable researchers to investigate details and help them delve deeper into a subject as they are semi-structured and require availability of time with the respondents (Gerring, 2007). Some of such interviews even lasted for 3–4 hours with some of the respondents in the field. In-depth face-to-face interviews were conducted for this study with the respondents ranging from the office-bearers of Parmarth like the secretary, state coordinators, district coordinators, project coordinators, team leaders to community mobilisers, volunteers, and community members. The officials shared their insights, knowledge, opinions, experiences, prospects, and challenges about the work during these interviews. They talked at length about all the initiatives the organisation was taking in association with the community members in different areas for addressing the issue of water scarcity during these interviews. Most of the interviews were conducted using interview guides containing semi-structured and open-ended questions. Questions for the interview guides were prepared based on the observations and understanding drawn from the pilot study carried out in the field and in accordance with the research objectives and questions. The preparation of the interview guide also required brainstorming sessions among the authors and hence was carefully thought through. The questions also varied for different respondents according to the nature of the roles and responsibilities they were undertaking. While conducting the in-depth interviews some other questions (beyond those contained in the interview guides) were also posed by the researchers to the respondents as and when they arose based on the responses of the interviewees to the previous questions. Clarifications were sought from the respondents by humbly requesting them to shed more light on some specific terms that they used during the interview.

The activists involved with a campaign being run by Parmarth on a national level were also interviewed as a part of this study. These interviews were instrumental in capturing the relevant stories, experiences, observations, and insights as to why they thought a campaign was needed for addressing the issues of water, what were the various activities that they were undertaking as a part of the campaign, and how successful they thought the campaign was. Most of the interviews lasted from 30 minutes to 1 hour and 30 minutes. The interviews were recorded electronically, and some important notes were taken immediately by the researchers as the conversation progressed, which on many occasions informed the field notes on that particular day. Later, all the interviews were transcribed, and the transcripts were used for drawing the relevant data insights for analysis in the research. A few stakeholders were interviewed multiple times and at different points in time during the research to capture their perspectives on new developments taking place in the initiatives that they were implementing.

In-depth interviews were mostly conducted with the key officials of Parmarth. However, there were activists and farmers associated with other water management initiatives and natural farming who were interviewed to gain some broad insights into such work in the Bundelkhand region. Their details are provided in Table 2.1 given below. Even though the number of recorded immersive in-depth interviews (for which the details are provided in table 2.1) with the key informants was 20, the authors had formally or informally interviewed at least 80 people during the fieldwork.

The authors engaged in formal and informal conversations with the key informants listed in the table above on several occasions during different field visits in order to seek updates about the progress of the projects and interventions. However, the dates mentioned in the table indicate the days when they were interviewed by appointment using the interview guide, and the interview was recorded. On other occasions, the researchers took notes of the interactions that took place on the go or while travelling together with the respondents to some places which could be considered as what Kusenbach (2003) calls "go-alongs." Kusenbach (2003) describes go-alongs as a "hybrid between participant observation and interviewing" which is very useful in "exploring the role of place in everyday lived experience" (p. 463). For her, "Go-alongs are more modest, but also a more systematic and outcome-oriented version of 'hanging out' with key informants – an ethnographic practice that is highly recommended in virtually all fieldwork manuals and textbooks" (Kusenbach, 2003, p. 463). Such "hanging out" with the key informants happened on several occasions during the fieldwork for this study, and this was of immense significance in developing a deeper understanding of the local settings as well as of the initiatives and interventions being explored. Carpiano (2009) believes that go-along is a method in field research which has a high utility in discovering the interactions between human beings and their environments. When it comes to data collection for a qualitative research study, go-alongs complement the other methods like observations, interviews, FGDs, etc. (Carpiano, 2009).

Table 2.1 List of in-depth interviews

S. No.	*Name and Designation*	*Profile of the interviewee*	*Place*
1.	Sanjay Singh	Secretary, Parmarth and National Convener, JJJA	Jhansi, UP
2.	Rupak Ghosh	State Coordinator, JJJA, MP	Chhatarpur, MP
3.	Rahul Singh	Project Coordinator, Integrated Watershed Management Programme (IWMP), Parmarth	Chhatarpur, MP
4.	Vishwanath Tiwari	Team Leader, Integrated Watershed Management Programme (IWMP), Parmarth	Chhatarpur, MP
5.	Ramesh Bhadoria	Team Leader, Integrated Watershed Management Programme (IWMP), Parmarth	Chhatarpur, MP
6.	Harishankar Sharma	Sarpanch, Baksohi Gram Panchayat, Chhatarpur, MP	Baksohi, Chhatarpur, MP
7.	Ravi Tomar	Ex-Project Coordinator, WASH (Water, Sanitation and Hygiene project funded by NABARD), Tikamgarh, Parmarth	Jhansi, UP
8.	Vinita Yadav	Cluster Coordinator, JJJA, Talbehat, Lalitpur, UP	Talbehat, Lalitpur, UP
9.	Rahul Singh	Team Leader, Integrated Watershed Management Programme (IWMP), Parmarth	Bijawar, Chhatarpur, MP
10.	Pushpendra Bhai	Convener, Apna Talaab Abhiyan (an initiative led by farmers of Banda to have ponds in their farmlands), Banda	Banda, UP
11.	Prem Singh	Organic Farmer, Banda	Banda, UP
12.	Sanjay Singh	Secretary, Parmarth and National Convener, JJJA	Jhansi, UP
13.	Rupak Ghosh	State Coordinator, JJJA, MP	Bada Malhara, Chhatarpur, MP
14.	Ajay Yadav	District Coordinator, JJJA, Chhatarpur, MP	Ghuwara, Chhatarpur, MP
15.	Manish Kumar	State Coordinator, JJJA, UP	Jhansi, UP
16.	Manvendra Singh	Project Coordinator, WADI, Parmarth	Lalitpur, UP
17.	Rajesh Kumar	District Coordinator, JJJA, Jhansi, UP	Babina, Jhansi, UP
18.	Shyam Bihari Gupta	Convener, Bundelkhand Kisaan Samriddhi Abhiyan, a campaign to promote natural farming among farmers of Bundelkhand	Jhansi, UP
19.	Vinod Bodhankar	Activist, JJJA and National Convener of Jal Biradari	Alwar, Rajasthan
20.	Ramendri	President, *Pani Panchayat*, Motran Village, Babina Block, Jhansi	Babina, Jhansi, UP
21.	Rajendra Singh	Waterman of India, Leader, JJJA	Ujjain, MP
22.	Sanjay Singh	Secretary, Parmarth and National Convener, JJJA	Ujjain, MP

Focus Group Discussions (FGDs)

Focus group discussions are data collection tools that are used by a researcher for drawing information out of a group conversation and are used increasingly in social science research these days (Morgan, 2008; Darlington & Scott, 2002). Barbour (2007, p. 2) describes focus groups discussions (FGDs) as an approach that relies on "generating and analysing interaction between participants, rather than asking the same question (or list of questions) to each group participant in turn, which would be the approach favoured by what is more commonly referred to as the 'group interview'." The aspect of FGD that makes it very relevant for this study is that it brings "together people who share a similar background… and create the opportunity for participants to engage in meaningful conversations about the topics that researchers wish to understand" (Morgan, 2008, p. 352).

During the fieldwork for this study, the authors conducted FGDs with the community members for whom the water management projects were implemented. The members of the *Pani Panchayat,* a women-led group constituted in villages for mobilising women to take action for addressing water-related problems, also participated in the FGDs. The FGDs were organised in the villages of the participants. The discussions were electronically audio-recorded, and simultaneously notes were taken for references that were helpful during the transcription of these FGDs and analysis of the data gathered through them. In addition to the recorded FGDs, on many occasions, the researchers interacted with community members in different villages and members of local groups. These interactions could not be recorded as they were brief or sometimes they were taking place as the authors or the respondents were on the go. But any meaningful stories or information that surfaced from such informal conversations, and related to the research questions of the study, were carefully written down in the diary or were recorded by the researchers on their mobile phones. Conversations took place with the participants of the recorded FGDs too, later, and anything worthwhile was included in the researchers' notes.

In the following chapters, some direct quotes of the participants drawn from the transcripts are used by the researchers to capture the language and impact of their articulation of a story or of a problem, in their exact words. However, the essence of these quotes is translated into English next to their quotes. In addition to the six formal (recorded) FGDs, the authors had several other group interactions without recording with the community members, water collectives' members, *Jal Sahelis*, farmers, and beneficiaries of the water management interventions.

Table 2.2 provides details of the recorded focus group discussions conducted during the fieldwork.

Table 2.2 List of focus group discussions (FGDs)

S. No.	*Participants*	*Initiative/Project*	*Place*
1.	Village-level Watershed Committee Members, Diha Village, Bijawar, Chhatarpur	Rajiv Gandhi Watershed Management Mission, Govt. of MP	Diha village, Bijawar Block, Chhatarpur, MP
2.	*Pani Panchayat* members and *Jal Sahelis*	Grassroots civil society building initiative by Parmarth supported by European Union	Kalothra village, Lalitpur, UP
3.	*Pani Panchayat* members and *Jal Sahelis*	Grassroots civil society building initiative by Parmarth supported by European Union	Kakradi village, Lalitpur, UP
4.	*Pani Panchayat* members and *Jal Sahelis*	Grassroots civil society building initiative by Parmarth supported by European Union	Bijrotha village, Lalitpur, UP
5.	*Pani Panchayat* members and *Jal Sahelis*	Grassroots civil society building initiative by Parmarth supported by European Union	Motto village, Lalitpur, UP
6.	*Pani Panchayat* members and *Jal Sahelis*	Grassroots civil society building initiative by Parmarth supported by European Union	Manpura village, Babina, Jhansi

Participant Observations

Historically, observation has been a significant method for conducting research of ethnographic nature in social science disciplines. It helps in effectively exploring what people do in a given setting in everyday interactions of their lives (Darlington & Scott, 2002). Participant and non-participant observations are two research strategies for a methodology employing an ethnographic approach (Gobo, 2008). Participant observation is one of the methodological approaches in qualitative research which is used in combination with other methods like document analysis, focus groups, and interviews (McKechnie, 2008). As Gobo (2008) puts it, participant observation enables a researcher to establish direct contact with the social actors as one stays in their local setting intending to observe their lives and behaviour by "participating in their everyday ceremonials and rituals" and "learning their code (or at least parts of it) in order to understand the meaning of their actions" (p. 5). In this study, the authors employ participant-observation methods to understand various events and activities being organised by Parmarth to draw insights and meanings (out of the rigorous field observational notes taken) which constituted a substantial part of the data collected for analysis.

The notes taken during the participant observation of the events were very useful in understanding the various aspects of the interventions and different stakeholders' engagement with them. The researchers had done participant observation of very locally organised events like village-level meetings of the organisation's officials and volunteers as well as the regional symposiums and the national-level programmes like national water conventions, consultation camps, etc. on the issues of water and other related matters.

A detailed account of all such events is provided in Table 2.3 below.

Table 2.3 List of events observed by the researcher

S. No.	*Event*	*Organised by*	*Venue*
1.	Consultation meeting at Arora village, Baldeogarh, Tikamgarh	Parmarth	Arora village, Tikamgarh, MP
2.	International Women's Day programme at Chhatarpur – Felicitation of *Jal Sahelis*	Parmarth	Chhatarpur, MP
3.	National-level Social Activists' Leadership Development Camp	Parmarth	Alwar, Rajasthan
4.	National Water Convention	*Jal Jan Jodo Abhiyan*	Vijayapura, Karnataka
5.	Night Choupal (a programme organised at villages which starts with Bhajans and folk singing followed by discussions on issues related to the village)	Parmarth	Nohra village, Babina, Jhansi, UP
6.	Natural Farming Training	Parmarth	Indian Grassland and Fodder Research Institute, Jhansi
7.	Fishermen Convention (organised annually on the occasion of Sharad Purnima for celebrating life and to mobilise the community)	Fishermen community of Chhatarpur	Jatashankar, Chhatarpur, MP
8.	Water awareness rallies and dialogue in various villages of Chhatarpur	Parmarth	Chhatarpur, MP
9.	Meeting of Parmarth officials with tribal population	Parmarth	Shekhar village, Babina block, Jhansi
10.	Farmers and Water Convention	Parmarth	Ghuwara, Chhatarpur, MP
11.	CSOs meeting for forming Bundelkhand Water Forum (an initiative aimed at defining roles for different CSOs to address drought-related issues)	Parmarth	Orchha, Tikamgarh, MP
12.	One-day natural farming workshop	Civil Society	Jhansi, UP
13.	Meeting at Parmarth office for developing strategies for immediate drought relief for farmers	Parmarth	Parmarth, Jhansi
14.	Social and Environmental Justice Contemplation Camp	Parmarth	Alwar, Rajasthan
15.	Water Conference, Bhopal	Parmarth	Bhopal, MP
16.	National Water Convention, Ujjain, MP	Parmarth	Ujjain, MP

In addition to the above, several meetings at the Jhansi office of Parmarth were also attended by the researchers with due consent of the organisation's secretary and other office-bearers during the fieldwork period. These included review meetings of specific projects, district coordinator's meetings, gatherings regarding preparations of some events or activities, and many others. These meetings were beneficial in understanding the communication processes within the organisation, between the officials at the management level, and the field staff. Similarly, the participant observation of the meetings and discussions by Parmarth's field workers with the community members in the villages gave deeper insights as to how they mobilised the communities for a particular intervention.

Document Analysis

Documents refer to the materials that are considered significant in understanding the historical context and the existing situation in a given setting and therefore, they form the basis for most of the qualitative research projects (Schensul, 2008). According to Prior (2008), "documents are viewed as conduits of communication between, say, a writer and a reader-conduits that contain meaningful messages" (p. 230). Therefore, a suitable way of analysing the documents is to focus on the content and the underlying meanings and stories within them (Prior, 2008).

The documents examined as a part of this study included the status reports produced by the NGO on water availability in the region, agriculture, migration, struggle of women and children for water, the resultant health-related risks (as mentioned in Parmarth's Evaluation Report of the "Establishing Women's First Right to Water Project" 2016), health of water bodies, etc., newsletters, evaluation reports of various projects, closure reports, projects' progress reports, annual reports, and others. Besides, other documents produced under the *Jal Jan Jodo Abhiyan* by the NGO, like the Ganga White Paper, Bheekhampura Declaration, Vijaypura Declaration, Sholapur Declaration, and Patna Declaration, were also analysed along with the minutes of stakeholder meetings for gaining deep insights into the efforts being made by activists from various parts of the country to protect water bodies and ensure access to water for all. While many of these documents were procured in hard copies, others were gathered in electronic forms. During the analysis, the authors identified relevant data from these documents to supplement the information under different themes of the study.

Fieldwork for the Study

Before the actual fieldwork for the selected case study began in August 2017, a pilot work in Madhya Pradesh, Uttar Pradesh, and Rajasthan states was carried out between February and April of 2017 which proved to be very useful in the selection of the case to be studied. During the pilot fieldwork that started on 22 February 2017, the researchers made contacts with the several organisations that were considered as potential cases for the study, initially. It was only after interacting with four non-governmental organisations (NGOs) – two in Madhya Pradesh, one each in Uttar Pradesh, and one in Rajasthan that the researchers decided to study the work of Parmarth after looking at the kinds of initiatives that they were undertaking in Bundelkhand. Following the pilot study, the researchers prepared a report based on the observations, interactions, and experiences from the visits to different places and organisations and discussed them with their peer groups and colleagues. The ensuing discussion helped in finalising Parmarth as the case study. The feasibility of doing research with various organisations in consideration was also taken into account that helped in zeroing in on this particular case study.

Once the case study was finalised, the data collection research tools, such as interviews and FGD guides, were prepared for the fieldwork. Consent forms and other mandatory documents for conducting interviews and FGDs were also prepared.

The case study for this research work started from Vijayapura, Bijapur, Karnataka, where Parmarth organised a three-day national water convention under its nationwide campaign, JJJA. During these three days, copious notes of the proceedings were taken during the different technical sessions and also interacted with various activists. They had gathered there from across the country on the call from Parmarth. It also enabled the authors to interact with the officials of Parmarth on the sidelines of the event. These interactions, especially with the office-bearers of Parmarth, helped a great deal in building rapport with them. These officials played the role of key informants as the research progressed through the fieldwork in the Bundelkhand region. One of the authors travelled with them by train from Bijapur to Jhansi after the national convention was over on 18 August 2017. The journey which lasted for more than 32 hours served as an opportunity to strengthen the relationship with the staff members of the NGO and some Bundelkhand-based activists working on water.

Later during the fieldwork, the authors met the state coordinators of Parmarth for Uttar Pradesh and Madhya Pradesh and sought information about the various projects related to water being carried out in both the states. During the first week of September 2017, the researchers got access to Parmarth's office library which helped in obtaining more information about all the work that they were doing in various parts of the Bundelkhand region. The initiatives and activities being undertaken in different parts of the country under their nationwide JJJA campaign were also analysed through the literature available in the library. The literature available in the library informed the researchers about Parmarth's previous and ongoing projects like – WASH, watershed development, grassroots civil society cadre building (formation of *Pani Panchayat* and identification of *Jal Sahelis* – discussed in a later chapter at length), WADI, etc.

During the regular visits to Parmarth's office in Jhansi, where the researchers were staying for the fieldwork in Bundelkhand, there was regular interaction with staff members associated with various projects and in-depth interviews were also conducted with them there. In-depth interviews were also carried out with staff members at different places in the region like Talbehat in Lalitpur district of Uttar Pradesh, Babina in Jhansi district of Uttar Pradesh, Bijawar in Chhatarpur district of Madhya Pradesh, Ghuwara in Chhatarpur district of Madhya Pradesh, Jatara in Tikamgarh district of Madhya Pradesh, and others where Parmarth had its local offices.

The focus group discussions with the members of the *Pani Panchayat* and *Jal Sahelis* were conducted in their respective villages. During these visits to the villages, the researchers also observed the work being carried out by these groups with the support of Parmarth. The villages included Kalothra, Kakradi, Motto, Vijrautha, Manpura, and others in different districts of Madhya Pradesh and Uttar Pradesh. During the fieldwork, the authors also attended *Ratri Choupal* (a programme organised in the villages at night, which starts with the singing of devotional and folk songs followed by discussions on the issues related to the village). Several such events and activities were attended by the authors at different points of time during the fieldwork including, fishermen convention at Jatashankar in Chhatarpur that lasted for the whole night, water awareness rallies, and dialogue in various villages of the Chhatarpur district of Madhya Pradesh for four consecutive days (out of eight days for which it lasted).

The researchers also travelled to more than 50 villages to understand the other projects related to water, environment, ecology, livelihood, etc. like the Kitchen Garden project, WADI project, Seed Bank initiatives, Grain Bank initiatives, natural farming efforts, and others. In addition to observing, experiencing, and understanding these initiatives in Bundelkhand, they also travelled to places such as (a) Tarun Bharat Sangh Ashram in the Alwar district of Rajasthan for various national-level programmes organised by Parmarth on two different occasions, (b) Bhopal for a Water Conference, and (c) Ujjain for a three-day national water convention to revive Kshipra river of the area and others.

In addition to this, the authors also travelled to Banda where an individual farmer named Prem Singh practices natural farming on his 10-acre land and also runs a school to train farmers in doing natural agriculture on their land. Prem Singh has set up a farming museum which traces the history of farming in India and displays various successful ways of doing natural farming without harming the health of the soil as well as the health of the human beings and the ecosystem. Besides, he has also set up a farming laboratory where he keeps experimenting with several kinds of manures for getting a good yield from the crops that he sows.

Coding and Analysis of the Data

The data for this research was gathered through various sources like interviews, FGDs, content analysis of the literature procured during the fieldwork, field notes, observations, etc. And the analysis process of this study began with what Yin (2011) calls a "careful and methodic organising of the original data" (p. 176) in the computer. This organising of data was followed by the compiling of the data. While analysing qualitative data, there is no universal approach or method to be followed (Yin, 2011; Darlington & Scott, 2002) and therefore, it is difficult to point out a specific time when the data analysis began for this study. In some form or the other, the analysis was also simultaneously happening when the data was being collected or when the in-depth interviews and FGDs were being transcribed as some sections of the transcripts were immediately copied and pasted into different files that were created to organise the data in line with the different theoretical frameworks used in this study.

In a qualitative research, the data collection and its analysis become iterative processes as the analysis of the data gathered at the early stage of the research offers insights for the further data collection process (ven den Hoonard & ven den Hoonard, 2008) as it happened with this research too. The analysis of the data collected during the pilot study helped inform the data collection processes for the full-fledged fieldwork. Later, as the work of the data collection progressed, followed by their analysis, the researchers kept realising the need for further data to be gathered during different stages of the fieldwork. The analysis and the interpretation of the collected data were not happening in isolation with the researchers alone. The consultations with colleagues and communication scholars at regular intervals were also a form of data analysis that not only informed the different phases of the data collection in the field but also assisted in shaping the outline of the book.

A careful transcribing of the in-depth interviews and the FGDs recorded during the different stages of the fieldwork was followed by the immersive reading of the transcripts for extracting the relevant quotes that captured the stories falling in line with the different thematic areas of the research. Reading and re-reading the transcripts were also beneficial for developing familiarity with the data. This was followed by the process of coding to derive categories and themes from the data. Coding refers to the process of drawing concepts from the available data through brainstorming so that the raw data could be opened up to all potentials and possibilities present in them (Corbin & Strauss, 2008). Benaquisto (2008) articulates coding as a process in qualitative research that helps in identifying, arranging, and systematically organising the ideas, concepts, and categories derived from the qualitative data. During the analysis of the data for this study, the researchers employed both open and focused or integrative coding. Open coding as suggested by Benaquisto (2008) is employed at the beginning of an analysis by closely reading the data for identifying as many ideas and concepts as possible; and at the later stage, focused coding helps in reconceptualising the various concepts and incorporating them into broader and more abstract categories.

Similarly, the observational data available in the forms of field notes and analytic memos were visited, revisited, analysed, and coded. And as the concepts and ideas emerged from them, they were linked with the different themes of the research and segregated to be included in thematic chapters of the text. The analyses of the data available in various forms as above were also carried out by referring to the scholarship available under different theoretical domains that had informed the topic of this research. This process of analysis was of great significance in locating the theoretical interests of the existing situations, change efforts, communication practices, and strategies of the given case study as they emerged through the data.

Bundelkhand: The Water-Stressed Region of Central India

Spread across a geographical area of 70,000 square kilometres, Bundelkhand covers 14 districts of Uttar Pradesh and Madhya Pradesh. It comprises seven districts of Uttar Pradesh, namely Jhansi, Lalitpur, Jalaun, Banda, Chitrakut, Hamirpur, and Mahoba, and seven districts of Madhya Pradesh, which are Tikamgarh, Chhatarpur, Sagar, Datia, Damoh, Panna, and Niwari.

Bundelkhand is a typical example of a semi-arid region that is characterised by unreliable rainfall where the rate of evaporation of water is higher than the rate of moisture received through precipitation ultimately resulting in unfertile and famine-prone land (Development Alternatives, 2012). Recurring droughts year after year have been a major point of concern for the Bundelkhand region for several decades. The region is prone to deficient rainfall and consequent drought due to both changing climatic conditions and man-made reasons (Singh, 2014). The water crisis in the region is so acute that it often leads to tensions between "water haves and have-nots" (Dasgupta, 2016).

The frequently recorded deficient monsoons have made this region vulnerable to drought, making the living conditions extremely harsh, especially in the rural areas where the people mainly depend on incomes from agriculture for sustenance. As it can be gauged from several news media reports, issues like farmers' suicide, starvation deaths, unemployment, large-scale migration, and abandonment of cattle have become common features of the region. Women and girls in various areas of the region have to travel several kilometres and spend hours to collect water for drinking and other purposes. In this dry and parched region, women and girls have been the worst victims of the water crisis. In many areas, when women have to travel several kilometres for fetching water, it makes their lives extremely vulnerable. In the rural areas across the Bundelkhand region, women and girls have to spend an average of 4–5 hours per day for securing around 20 litres of drinking water (NIDM, 2014). Around 25 per cent of the population in Bundelkhand belongs to scheduled castes out of which around 80 per cent lives in rural areas which witness feudal social structure as can be gauged from the rampant social and gender discrimination prevalent here.

From 2012 to 2017, Bundelkhand had seen six droughts except for the year 2016 when it received close-to-normal rainfall. In the last 16 years, the region has faced 13 droughts. If we go by the reports from different agencies, the situation was not like this always in the region. According to a *Down to Earth* report (Jitendra, 2016), the region faced drought every 16 years in the 18th and 19th centuries and this frequency increased to every five years between 1968 and 1992. In 2017, the deficiency in the rainfall was as high as 59 per cent (Chauhan, 2017) that caused a high level of distress and hopelessness among the farmers of the region forcing them to migrate to other cities in search of livelihood. Mahapatra (2017) reveals that the Bundelkhand region of both Uttar Pradesh and Madhya Pradesh was experiencing 15 consecutive crop losses in the year 2017 thanks to the changing climate and also warned that such situations were going to be the new normal in the days to come.

Demographically, the total population of the Bundelkhand region according to the Census data of 2011 is around 18.3 million out of which the seven districts of Uttar Pradesh have a population of around 9.6 million while seven districts of Madhya Pradesh are home to around 8.6 million population.[1] The sex ratio in Uttar Pradesh's Bundelkhand region is 877 and in Madhya Pradesh part of the region it stands at 930 both being lower than the national average of 940. The census data revealed that around 78 per cent of the population of the region are living in rural areas.[2] The rural areas of the Bundelkhand region are marked by a low density of population; however, the plain areas of the region like Banda, Jalaun, Hamirpur, and Jhansi are marked by a higher density of population. If we look at the average population density in the region, the seven districts of Uttar Pradesh account for 277 per square kilometre while seven districts of Madhya Pradesh coming under the region have an average population density of 233 persons per square kilometre.

In Bundelkhand, even as more than 75 per cent of the population depends on income from agricultural activities and 96 per cent of the income is generated from agriculture and livestock, the conditions are continuously turning unfavourable for the farmers with the fluctuating weather and increasing wasteland (An Inter-Agency Initiative, 2017). With 60 per cent of the agricultural land in the region being rain-fed, the frequent droughts have led to a situation where farmers have to depend on a single crop that they sow during the rabi season with whatever rain they receive during the monsoon.

A peep into the region's history reveals that it was once very efficient in water harvesting and management. Bundelkhand being historically drought-prone, some of the dynasties like Bundelas and Chandelas had built huge networks of ponds, lakes, tanks, step-wells (*bawdies*), wells, etc. more than one thousand years ago, to drought-proof the region by minimising the surface run-off (Perspectives, 2010). These water harvesting structures made the region drought-proof by recharging the groundwater reserves resulting in the availability of water in wells and step-wells and maintaining the moisture in the soil.

According to Parmarth officials, every village in Bundelkhand used to have ponds as the topography of the region was favourable for creating ponds for water harvesting and conservation. There would be different kinds of ponds, with different kinds of usage – like some would be used for collecting rainwater, some for irrigation, and others for making drinking water available to animals, etc.

According to a *Down to Earth* report (Jitendra, 2016), the decline of these water management structures started around five decades back, i.e. from the 1960s when land consolidation laws (chakbandi) were amended in various states. Earlier, landholdings were developed keeping in mind the water resources ensuring that every farmer had water available for irrigation. But, according to the changed laws landholdings were designed in rectangular shapes that did not take into consideration the aspect of water availability. Besides, the relentless and haphazard construction work in and around the urban settlements led to a large-scale encroachment of the traditional water conservation resources that blocked the channels through which these water harvesting infrastructures were replenished. Even as the Bundelkhand region is home to several prominent rivers like Yamuna, Ken, Betwa, Bagahin, Pahuj, Chandrawal, Kali Sindh, Chambal, and others, the health of these rivers is fast deteriorating due to human activities like deforestation, pollution, and daming. Therefore, in the face of climate change and the relentless human activities affecting the ecology in a negative way, water, the most precious natural resource for sustaining lives, is becoming scarcer in the region.

Parmarth Samaj Sevi Sansthan: An Overview

Parmarth Samaj Sevi Sansthan (PSSS), which is the case study for this research, is a non-governmental and non-profit organisation registered under the Society Registration Act 1860. The organisation was also registered under the Foreign Contribution Regulation Act (FCRA) in the year 2000. Established in the year 1996, the NGO undertook projects and interventions that were mainly aimed at bringing qualitative changes in the lives of the poor and deprived sections of the society. According to the secretary of Parmarth, Sanjay Singh, the organisation was formed by a group of youth who were willing to make their contribution to bringing about a change in this backward region. He said that the objective of the organisation was to provide sustainable ways of livelihood generation and work towards the socio-economic and political development of the region. Even today, Parmarth is working towards developing the semi-arid drought-prone region of Bundelkhand in a sustainable way in association with the local people and government and administration.

The working philosophy of Parmarth is based on the approach of community empowerment and collective action through capacity building programmes, facilitating participation, seeking accountability for rights and entitlements, and ensuring transparency in its day-to-day functioning.[3] The strength of Parmarth, according to its officials, lies in its collective mobilisation of civil society to build a sustainable model of development at local, regional, and national levels. Following these principles, as it was also observed during the fieldwork, the organisation has worked towards formation of community-based organisations (CBOs) and self-help groups (SHGs) at local levels for strengthening the grassroots cadres of civil society. These CBOs and SHGs include village-level watershed committees, *Pani Panchayat* groups, *Bal Samuh,* and others. In addition to these local-level groups and organisations, it has also worked towards formation of a network at district and state-level like Pahuj Vikas Manch, Aapda Nivaran Manch, Bundelkhand Water Forum, and others. These forums and platforms work towards providing livelihood assistance to the poor and marginalised sections of society by facilitating public hearings with the representatives of the government and the local administration. Such activities are aimed at seeking accountability for the rights of the people belonging to the poor and vulnerable sections of society. These forums also work for advocacy and policy planning for drought-related problems at state levels in both Madhya Pradesh and Uttar Pradesh with their respective governments.

The major areas of Parmarth's interventions are[4]:

- Natural resource management and water conservation
- Livelihood and sustainable agriculture
- Climate change
- Water sanitation and hygiene (WASH)
- Child rights protection
- Education and vocational training with special focus on adolescent girls
- Women empowerment
- Health and nutrition
- Disaster reduction response
- Establishing women's first right to water resources
- Integrated water resource management (IWRM)
- Integrated tribal development fund (ITDF) – WADI
- Watershed Development Fund (WDF)
- Village Development Programme (VDP)
- Uttar Pradesh Participatory Forest Management & Poverty Alleviation Project
- Targeted interventions
- Umang Community Care Center (UCCC)

- Beti Bachao Abhiyan (save the girl child campaign)
- Kasturba Gandhi Aawasiya Balika Vidyalaya (KGAV)

Out of all these different initiatives and interventions being undertaken by Parmarth, this study focuses on its initiatives like "Establishing women's first right to water resources," "Livelihood and sustainable agriculture," "IWRM," "ITDF-WADI" project, "Health and nutrition," and "WASH."

Under its project to establish women's first right to water resources, Parmarth has formed groups of women in 96 villages of 60 Gram Panchayats of three districts of Uttar Pradesh's region of Bundelkhand, namely Jalaun, Hamirpur, and Lalitpur. The project aims at ensuring drinking water security for all in the villages through an integrated water resource management approach. The project aims at ensuring a participatory decision-making process at Panchayati Raj Institutions (PRIs) through women's advocacy groups and their leadership. One of the project objectives is also to promote women's role in the decision-making processes pertaining to the management of community water resources.

Through the livelihood and sustainable agriculture initiative, Parmarth, with other farmers' groups and agriculture institutions, is working towards promoting natural farming practices among the farmers as a response to the climate change phenomenon and the recurring droughts in the region. Through trainings and awareness programmes, it is educating the farmers to use their agricultural lands in ways that help to conserve water during the monsoon. They also train them to use new and cheap techniques to irrigate their farmlands, sow less water-intensive crops, and shun the practice of using excessive chemical fertilisers and use locally made manures and yet get better yields. The researchers attended some training programmes to understand these efforts and also made visits to the farmlands of the farmers who have adopted these practices in farming.

Under the integrated water resource management (IWRM) project, Parmarth has worked with the local communities in around 40 villages of Chhatarpur and Tikamgarh districts of Madhya Pradesh for watershed development. And through this project, various check-dams, stop-dams, and farm-ponds have been developed in the villages with financial support from the Government of Madhya Pradesh under its programme titled Rajiv Gandhi Watershed Management Mission (RGWMM). The work of developing these water management structures was carried out by forming a village-level watershed committee (VWC) which would monitor these works and ensure quality and success.

Under the ITDF-WADI project, Parmarth has helped landless families from the tribal communities of the Talbehat block of Lalitpur in Uttar Pradesh to develop fruit orchards in their farmlands to help them develop a sustainable livelihood source. Around 1000 families from the Sahariya tribe in this area were identified who benefited from the programme.

As a part of its health and nutrition programme, Parmarth has worked for enabling poor families in the rural areas of Bundelkhand to develop kitchen gardens in their backyards to grow vegetables so that they could grow some nutritional food of their own. Under the WASH project, the NGO has worked towards making some villages of the region open-defecation free (ODF) with constant efforts for mobilising the local community members, especially women in the rural areas. All these initiatives are studied and discussed at length in the later chapters of the book.

From the beginning of this research to carry out study on grassroots water management interventions from an environmental communication lens, to the present, when it has culminated in the stage of writing, the journey has been highly enriching. Every stage of this study from literature review to data collection, then the analysis and writing helped the authors learn, unlearn, reflect, and build upon the understandings gathered from the field, data, consultations with colleagues and friends as well as the available scholarship from the field. And these insights were crucial in terms of guiding the researchers in determining the thematic areas and structuring this text. The following three chapters present the discussions and analyses around the three main thematic areas identified for this research, viz. "theory of change," "environmental communication," and "civil society" while the concluding chapter sums up the study by highlighting the key findings from the thematic chapters and exploring the possibilities of some specialised areas of research within the field of environmental communication with special focus on water and sustainable development.

Notes

1 "Bundelkhand's population and literacy chart" retrieved from – https://bundelkhand.in/ict/bundelkhands-population-and-literacy-rate-chart-via-census-data-2011 on 6 February 2020.
2 "Changes in Demographic Characteristics in Bundelkhand Region: Evidences from Census 2001 & 2011" retrieved from – https://bundelkhand.in/changes-in-demographic-characteristics-in-bundelkhand-regionon on 6 February 2020.
3 Parmarth's annual report 2016–17.
4 Parmarth's Annual Reports for 2012–13 and 2016–17.

3 Building Capacity as a Theory of Change

The Making of *Jal Sahelis*

In the area of development and social change, non-governmental organisations (NGOs), in the last few decades, have emerged as significant players. The increasing role of NGOs in bringing about change in collaboration with the different stakeholders has not only been recognised by the governments in different countries but also by international organisations having global operations like the United Nations. Especially when it comes to the concerns associated with the sustainability of the earth's resources, NGOs are considered as one of the important actors in attaining sustainable development goals (SDGs).[1]

The 1992 Earth Summit held in Rio de Janeiro has explicitly sought the commitment and active involvement of the non-state actors in the sustainable development blueprint prepared during the event (Gemmil & Bamidele-Izu, 2002). NGOs are considered as agents that bridge the gap between government or international organisations and the communities at grassroots level. They have trained staff who are capable of informing, educating, mobilising, and organising people in remote areas where government officials and professionals of international agencies find it difficult to reach (Joshi & Kumar, 2007). As the concerns of environmental degradation, climate change, and associated distress with these phenomena continue to grow globally, the role of NGOs and other civil society groups seems to gain even more significance, considering that their local interventions could have global relevance (Samuel & Thanikachalam, 2003). The issue of climate change has led to changing weather patterns and increased the frequency of extreme weather events (Anderson, 2014). Thousands of lives are lost owing to extreme weather events globally, and millions suffer due to harsh weather conditions like droughts, floods, hailstorms, etc.

India too faces such weather conditions every year in different regions. The Bundelkhand region situated in parts of Uttar Pradesh and Madhya Pradesh provinces of India is one such area that witnesses recurring droughts year after year due to low precipitation. This causes large-scale migration and immense distress among the people inhabiting the region. People struggle on a daily basis for obtaining drinking water, as well as water for everyday chores. Women and girls spend several hours, ranging from 2 to 8 hours every day, for fetching water from distant places.

DOI: 10.4324/9781003350286-3

Looking at the harsh living conditions of the people of Bundelkhand due to the scarcity of water, there are several civil society organisations, NGOs, and voluntary organisations that have taken up the task to address the issue. The governments – both central and state – allocate special funds for Bundelkhand in their budgets to help the region cope with the issue of water crisis in an efficient manner. International organisations like the UN, Development Alternatives, European Union, Water Aid, Action Aid, Welt Hunger Hilfe (WHH), Deutsche Gesellschaft fur International Zusammenarbeit (GIZ), etc. too have several projects running in collaboration with local NGOs to help people cope with the issues of water and associated livelihood concerns. Different NGOs and their donor agencies apply different methods and approaches to find solutions to the existing problem of water and livelihood crises.

Many a times, when NGOs, voluntary or civil society organisations, come forward and attempt to work for tackling a particular issue in collaboration with the local communities, their initiatives and ideas motivate the communities to get involved with a purpose to make a difference in their own lives as well as the lives of their fellow community members. However, for the organisation spearheading the initiative to bring about a change, it's challenging to channelise their energies in an effective way so that the desired goal could be achieved. That's where a good theory needs to be developed to help in making effective strategies. And the key idea that surfaces from the way community involvement is sought for the effectiveness of the change-making efforts is the concept of "participatory communication" which, according to Manyozo (2012), serves as a significant strategy for stakeholder engagement through "dialogue, respect for local knowledge and collective decision-making" (p. 155). This chapter studies Parmarth's theory of change and the integrated participatory approach that enable community members to become committed partners in interventions of different nature related to water. And in this light, it becomes imperative to first understand what a theory of change is, how it came into being, how it became an integral part of the change initiatives and processes, and what are the ways in which it is being used by the NGOs?

This chapter is divided into eight sections. The one after this section defines the concept of theory of change by referring to the work of scholars and practitioners from the field of development and communication. The third section of the chapter presents the overall situation of water crisis in the Bundelkhand region and the associated struggles of women in their everyday lives. The fourth section maps the social norms prevalent in the region and how Parmarth works towards changing those norms to enable women to deal with the hardships created due to the unavailability of water in their vicinity. The fifth section describes the process of formation of the women water collectives in the villages called *Pani Panchayat* and the identification of *Jal Sahelis*, women who lead these collectives at the village levels. The sixth segment of the chapter discusses how the formation of these women's groups leads to the participation of the community members in the water management initiatives and the resultant empowerment of the women in exerting their control and ownership in a change-making process. While the seventh section explains how the water management projects are made sustainable thanks to the active involvement of the women collectives in the villages, the eighth part of the chapter, i.e. conclusion, pulls together the important insights from the above sections and analyses them in the context of theory of change scholarship.

Theory of Change: Defining Pathways and Outcomes

> "Those who are in love with practice without knowledge are like the sailor who gets into a ship without rudder or compass and who never can be certain whither he is going."
>
> **– Leonardo da Vinci**[2]

In the last one decade or so, the development sector has seen a lot of funding agencies and implementing organisations shift their focus from conventional project-based approaches to the "Theory of Change" thinking in order to understand better how change happens through a particular initiative in a given area of intervention (James, 2013; Reeler & Blerk, 2017). Theory of change has come to be accepted as an alternative to the logical framework popularly known as the logframe approach as the latter was used mostly in the business world as a project planning tool which was oblivious of its own limitations in terms of assumptions about the desired change whereas the former is useful in identifying issues and bringing change even in complex and unpredictable situations (Reeler & Blerk, 2017). While the theory of change approach helps a development organisation see the bigger picture by taking into account complex realities, the underlying assumption within the logframe approach is that merely a well-planned project along with a good vision will bring about the desired change without considering the situations on the ground. Therefore, the theory of change is increasingly being used, these days, by donors as well as implementers so as to explicitly envision long-term goals and determine the pathways for achieving them. A clearly specified theory of change has become an essential component to be included in funding applications these days. As such, even the NGOs which are working in remote areas of the world that have donors supporting their work tend to develop their theories for influencing change in the areas of their operations. The NGOs working in the Bundelkhand region are no exceptions.

In simple words theory of change could be defined as a process that seeks to develop what Rick Davies explains as "The description of a sequence of events that is expected to lead to a particular desired outcome."[3] James (2011) sees theory of change as, "an ongoing process of reflection to explore change and how it happens – and what that means for the part organisations play in a particular context, sector and/or group of people" (p. ii). James (2011) believes that theory of change helps organisations ask important questions to themselves about the work they are doing, to develop an effective framework for monitoring and evaluation, obtain clarity about their approach, and at the same time also strengthen partnerships and communication strategies. As cited by Vogel (2012), Patricia Rogers, a noted evaluation expert, defines theory of change as a way to articulate "underlying assumptions about how change will happen in a programme" (p. 5). She considers it as an "on-going process of discussion-based analysis and learning that produces powerful insights to support programme design, strategy, implementation, evaluation and impact assessment, communicated through diagrams and narratives which are updated at regular intervals" (Vogel, 2012, p. 5).

As viewed by some other scholars, theory of change is a process that outlines the range of activities needed for achieving certain desired outcomes (Allen et al., 2017). With these descriptions of the theory of change by different scholars, it can be said that the aspect of social change is at the heart of the theory of change process and approach. Anderson (2005) believes that the theory of change is "a pathway of change that illustrates the relationship between variety of outcomes that are each thought of as preconditions of the long-term goal" (p. 7). She further says that theory of change is also helpful for the stakeholders of a programme to develop a shared understanding of the existing issues and how they are going to address them by ensuring clarity for everyone involved in the project. For any intervention aimed at bringing about a change, the theory of change approach suggests it has to begin by assessing the prevailing situation in a given context and then by identifying and bringing the stakeholders and actors together.[4] This chapter outlines and analyses the different steps involved in Parmarth's intervention to address the issue of water scarcity in different parts of Bundelkhand. And the first step that it takes for a change process is assessing the situation. It also discusses how the theory of change approach enables Parmarth to develop effective communication strategies to engage with different stakeholders at different stages of an initiative. And also finds resonance with the "formation of reliable communication networks" as mentioned by Manyozo (2012) with the coming together of different stakeholders that facilitate an atmosphere of mutual exchange of thoughts and ideas among them.

Situation Analysis: Water Crisis and Women's Struggles in Bundelkhand

To capture the existing realities at the ground in terms of water availability and the related struggles, Parmarth, as informed by its state coordinator for Uttar Pradesh, Manish Kumar, conducted a baseline survey using participatory rural appraisal (PRA) tools in three districts of Uttar Pradesh coming under

Bundelkhand region namely – Jalaun, Lalitpur, and Hamirpur in the year 2011. He said it was a part of a project called, "Ensuring Women's First Right to Water" which was financially supported by the European Union based on the proposal of Parmarth. The survey revealed that most of the households faced drinking water crisis as well as a shortage of water for irrigation and other household activities. The severity of the scarcity of water was such that the women had to travel several kilometres every day to procure water in most of the villages. The farmers were not able to sow crops in their agricultural fields. Young girls were not able to go to schools as they had to spend hours collecting water for drinking and other household chores. The cattle were abandoned to fend for themselves. Kumar said that such a situation emerged in Bundelkhand because of the apathy of the government and administration to preserve the old water management systems and the lack of awareness among the local communities about the deteriorating water conservation structures. No attention was paid towards developing new infrastructure for water harvesting that could lead to recharging the aquifer and the water table of the region. Even as some structures were developed by the government for water conservation like check dams and stop dams, most of them got damaged due to poor quality of material used in them because of the rampant corruption.

The agricultural productivity completely depends on the availability of water as well as the judicious use of the limited water and management of the water resources. Therefore, in the face of climate change and the relentless human activities affecting the ecology in a negative way, water, the most precious natural resource for sustaining lives, is becoming scarcer in the region. While the major parts of urban areas get piped water supply from the municipal bodies, the rural areas face crisis-like situation, especially during the summers. As the summer approaches and agricultural farmlands go uncultivated due to the want of water, men start to migrate to cities like Etawah, Mainpuri, Lucknow, and Kanpur to work in the brick kilns or to the bigger cities like Delhi, Mumbai, and others where they work as construction labourers. While men migrate from their families in search of livelihood, women have to stay in the village to take care of the children and the elderly members of the family. They are the ones who have to struggle for water on a daily basis. In Bundelkhand, it is the women who completely shoulder the responsibility of fetching water as it was visible during the visits to numerous villages wherein women could be commonly sighted carrying water pots on their heads and waists from some of the functional hand pumps and the wells. Young and adolescent girls could be easily seen carrying water on their heads commuting long distances and sometimes also riding bicycles with two huge water containers hanging with the carrier. While fulfilling the duties of fetching water, women do not only pay a heavy price in terms of the hours that they spend every day ensuring the availability of water for drinking and other chores but also in terms of health, money, and dignity. They become vulnerable to incidences of eve-teasing and sexual harassment as they travel alone for water in remote areas.

Being women and poor in the region, they never feature in the scheme of things of the government and administration when they make development-related policies, points Sanjay Singh, the secretary of Parmarth. He feels that the region is meted out with the treatment of "administrative apathy." And if the women are from the lower castes or tribes, their struggle is more complex, he added. The months from April to July are worst for the women of the rural areas of the region in terms of the painstaking efforts that they have to make on a daily basis to procure water.

Vineeta Yadav, the team leader of Parmarth in the Talbehat block of the Lalitpur district of Uttar Pradesh, said such is the condition of the women in the villages that they have no respite from the heavy burden of fetching water even when they are ill. Even during pregnancy, women could be seen carrying water containers as there is no one to relieve them of this unfortunate duty, she added. The scarcity of water in the villages not only takes a toll on the lives and health of the women, but the young girls of the families too have to pay a heavy price as they cannot go to schools in the process of assisting their mothers and families in collecting water for the household. And Bundelkhand being rich in livestock, it's not just the family for which they have to collect water, in most of the homes, they have to arrange water for the cattle too. This is evident from the low female literacy rates of the three districts of Uttar Pradesh, i.e. Jalaun, Hamirpur, and Lalitpur coming under the Bundelkhand region which are 49.21 per cent, 40.14 per cent, and 32.97 per cent respectively according to UNDP (2007).

Manvendra Singh, one of the project coordinators of Parmarth based in the Talbehat block of the Lalitpur district, informed that in a survey conducted by Parmarth it was found that women in several villages of the block had to spend half-an-hour to one-and-a-half hours at one go for collecting water for drinking and other household purposes. He said that most of the villagers have buffaloes in their households and one buffalo requires at least 30–35 litres of water every day for consumption. So, an equal amount of time needs to be devoted by the women for arranging water for the livestock as they do for their families which add to their physical burdens.

The UP state coordinator for Parmarth, Manish Kumar, said that according to a survey that they conducted in 131 villages of the region, merely 7 per cent had drinking water available to them for the entire year while more than 60 per cent villages had drinking water availability for only one month. In 31 per cent of them, it's available only for two months. During the summer season, 80 per cent of the hand pumps go dry in rural areas which are otherwise the chief source of obtaining water. Kumar said that Parmarth has been working on the issues of water and women empowerment for over one-and-a-half decades and has contributed to the revival of the Pahuj River in Bundelkhand and many other traditional lakes and ponds. However, looking at the plight of the women in the region in terms of their struggle for water, it decided to carry out some work which could help their cause. "So, first of all it was important to understand the social structure, societal norms and practices so that interventions could be designed that would address the issue in a particular social setting," he said. Therefore, the first thing that Parmarth did was to carry out a baseline assessment to understand the caste hierarchies, decision-making processes within the family structures, and the social norms that played out in the family and the society.

Understanding the Social Norms: Feudalism and Patriarchy Driving Social Behaviours

Sound understanding of the social norms in a given society could help a development organisation ask essential questions as to why people do certain things and identify the "causes and drivers of behaviour" (Bicchieri & Noah, 2017; Bicchieri, 2006; Bicchieri, 2017; Chung & Rimal, 2016; Vaitla et al., 2017). Scholars from various disciplines have established the idea, with empirical evidence, that social norms have a profound influence on the behaviour of individuals and social groups in a particular context and setting (Yanovitzky & Rimal, 2006). Understanding the influence of social norms on people's behaviours and attitudes is significant while trying to study human interaction processes, and therefore, there is constant building-up of scholarship in social sciences that focuses on the domain of social norms (Chung & Rimal, 2016).

For Chung and Rimal (2016), "norms are rules of behaviour that exist at both formal and informal levels" where the informal norms are considered to be more reinforcing and salient (p. 5). Bicchieri and Noah (2017) outline two kinds of expectations – empirical and normative – as parts of social norms that individuals and groups have to conform to. They say empirical expectations are those that the individuals in a reference group conform to thinking that others in the group are doing so while the normative expectations are those that most people in a network feel that they should follow (Bicchieri & Noah, 2017). For organisations to carry out successful interventions for bringing about change in the existing social norms and introduce new ones, it becomes imperative to examine the drivers of the behaviours and attitudes and, hence, identify empirical and normative expectations, social customs, sanctions norms, etc. (Bicchieri & Noah, 2017; Chung & Rimal, 2016; Yanovitzky & Rimal, 2006).

Tankard and Paluck (2016) discussed the power of influential individuals in creating new social norms who are early adopters of new ideas and practices as outlined by E.M. Rogers in his work on the diffusion of innovations. Such early adopters have a significant role to play in defying the social norms existing for ages in various rural settings of the Bundelkhand region of India where women have largely been pushed to the margins due to the feudal and patriarchal systems and mindsets of the society.

Parmarth realised that it was the existing norms in the region that created a divide between men and women when it came to determining the responsibilities for the family and decision-making about natural resources.

A closer look at the situation on grassroots level divulges that it was women who were at the receiving end due to societal norms. Women in many rural settings of Bundelkhand spending 4–5 hours every day and walking miles for fetching water is not unusual or strange as it is a widely reported fact (An Inter-Agency Initiative, 2017; Development Alternatives, 2015; NCW, 2005; NIDM, 2014; Water Aid, 2008). In some areas, it is even 7–8 hours which could be utterly unimaginable for people living in urban areas. Interacting with the women in the remote villages of the Bundelkhand region reveals the harsh realities that they have to live with due to the dying water bodies and changing climate. For rural women, the struggle for procuring water begins as early as they get up in the wee hours of the day and goes on till the end of the day. As Bicchieri (2006) puts it, "the very existence of social norm depends on a sufficient number of people believing that it exists and pertains to a given type of situation, and expecting that enough other people are following it in those kinds of situations" (p. 2), all women similarly followed the practice of procuring water for their household looking at the others in their community and considering it a norm. Citing the self-categorisation theory,[5] Chung and Rimal (2016) state that an individual is more likely to conform to the social norms of a group when one identifies oneself with that group, and this is what is reflected in the lives of these women and men of numerous villages in Bundelkhand as for them the social acceptance in their communities is of paramount importance. Thus, for them to not comply with the existing customs and practices is beyond their imagination.

Even as gender equality is one of the central aspects of the Sustainable Development Goals (SDGs) as well as the Millennium Development Goals (MDGs), gender discrimination is still widespread in the rural hinterlands of the country. The deep-rooted patriarchal mindset and practices still dominate the rural way of living. According to a report by UNICEF (2007) on the state of women and children, women and girls have to face discrimination in every walk of life be it access to resources, education, opportunities, political power, etc. The discrimination is meted out in various forms like gender-based violence, oppression of girls and women, limited personal and professional choices, and denial of basic human rights, the report revealed. It observes that to tackle the issue of gender discrimination and to ensure children's rights, it is crucial to make sure that both boys and girls get equal opportunities for getting an education. However, access to education for many of the girls in the rural areas of Bundelkhand is a distant dream as they are supposed to share the burden of the household responsibilities of the mother which may vary from cooking, washing utensils, and cleaning home to fetching water or taking care of the younger siblings when their mothers go to collect water from outside. For the families in rural Bundelkhand, there is a popular belief that even if the girl children are educated, they are not going to get any economic benefits to the family as they have to get married and go to another family.

The secretary of Parmarth, Sanjay Singh, during one of the in-depth interviews said that for the uneducated communities residing in the villages, daughters are liabilities as a result of which they marry them as soon as they reach 14–15 years of age. According to a study carried out by Parmarth and Women's Feature Service in 2011, in Lalitpur district of Uttar Pradesh one of the project districts of Parmarth, 80 per cent of the girls get married before they attain the age of 18 as they are considered "*parayadhan*" (someone else's wealth) socially and culturally as they belong to the families which they are married into.[6] The study further underlines that over the ages, the general mindsets have been shaped in such a way that even for the officials and authorities at the government departments and local governing bodies, infrastructure is meant to be gender neutral. For them, "Pipes are pipes and taps are taps, for both men and women – that is the common sense," but in reality even the "design and location of a hand pump can have profound consequences for women" (p. 20). Singh interestingly observed that in Bundelkhand, the problems exist not only because of the poverty in terms of resources and wealth of the households but because of the "poverty" in terms of vision in policy-making because "it's not just about hand-pumps and taps but about the lives and health of the women."

Manish Kumar quipped that in a state like Uttar Pradesh and in one of its most backward regions, i.e. Bundelkhand, gender and caste identities still have a massive impact in the everyday lives of the people. Here the feudal and patriarchal culture and practices still dictate the terms in day-to-day affairs of the people in both public as well as private spaces; he pointed out that

> women have to cope with the issue of gender inequality in their everyday lives in such a way that they do not even realise that they are becoming a victim of the age-old traditions and beliefs as for them it is common societal norms that they have to follow being women and girls.

Vineeta Yadav, the team leader of the project, "Establishing Women's First Right to Water" based at Talbehat block of Lalitpur district recalled how challenging it was for her to be able to talk to the women in the villages of the region during the beginning of the project as they would all be shy talking to the outsiders and remain head bowed with their faces covered with the *pallus* (veil) up to the neck. She said that for them it was a tradition to stay covered in their in-laws' house especially before the father-in-law or other senior male members of the family and while moving out. Yadav informed that the custom of covering the faces with *pallu* or *ghunghat* (veil) as they call it locally was not only among the communities belonging to the scheduled castes but also among upper castes. "Most of the women here faced similar kind of treatment in their in-laws' house no matter which caste they came from," she underlined. She said that during the beginning of the project in 2011, the women would hardly open their mouths when we would try to talk to them about their everyday experiences of life, their problems, issues related to water, etc. if there were any male members of the families around. However, when there were no male members, they would open up freely and narrate their stories about the everyday challenges that they would face with no water source being present at their houses, she recollected. According to Yadav, the feeling of being an inferior gender was so deeply entrenched in the minds of the women that they would have to seek permission from their husbands or other elderly male members of the family in the absence of their husbands even for going to the market for purchasing their personal items.

The existing social norms in rural areas of Bundelkhand have, however, assigned women with the responsibility of collecting water for the household to run. No matter how far the water resources are from their houses and how much time they have to devote, this responsibility has to be taken up by them. One of the *Jal Sahelis* of Kakradi village of Lalitpur, Kalan Devi said,

> We would have no choice but to concede in front of the family traditions and will spend hours for collecting water from the distantly located water resources even if our bodies were burning with high fever or any other ailments. Being born in this part of the world looked like it was our destiny to silently adhere to the existing norms and curse our fortunes. We would think that if only our water-related worries were taken care of, we would be in a much better position.

Team leader Yadav said, besides collecting water women also have to make sure that the house is clean, cattle have water to drink, and in many cases have to work in the field as well, but all these tasks that they do to keep their houses running do not have any social or economic significance to the families or the society. The UP state coordinator said that even as dairy is considered one of the economic activities contributing to the state's gross domestic product (GDP), women's contribution in looking after cattle is not considered to be an economic activity. Therefore, he said changing the situation at the ground level required change in the existing social norms, assessing the gaps between the issues facing the rural communities and the institutional mechanisms and bridging them by bringing different stakeholders together to deal with the struggles related to water.

Women as Stakeholders: Formation of *Pani Panchayat* Groups

Kumar, the UP state coordinator, said that once they had gathered a substantial understanding of the customs, traditions, family hierarchy, and the social norms underlying women's struggle for water and conflicts through their baseline surveys in three districts, they realised that the issues of water could not be addressed without bringing various stakeholders together, i.e.

– women, communities, local governing bodies like Panchayats and blocks, district administration, etc. The goal of ensuring water for all sections of the society could be achieved only with the greater participation of various stakeholders in their different capacities, especially in the areas which faced severe water shortages, he said.

Manvendra Singh, the project coordinator for the project "Ensuring Women's First Right to Water" based at the Talbehat block of the Lalitpur district, says,

> having realised that women were the worst victims of the feudal and patriarchal system, the customary norms, and the situations of the water crisis, we decided to first reach out to them and mobilise them to act towards finding a solution to their everyday struggle for water. We thought of talking to them about their *adhikar* (rights) and *samman* (dignity) so that they could feel strong and muster the courage for fighting for their rights and entitlements.

But that was not going to be easy as they were accustomed to a particular kind of lifestyle and societal customs that were ingrained in their way of thinking and conducting themselves. However, Singh said, "We knew that the need was to reach out to them on a regular basis and keep trying to convince them in the right manner and one day they will rise to assert their voice." On being asked what gave them this kind of confidence, he said that they had it in their mind that the women who were capable of silently coping with the extremely stressful and strenuous situations every day were indeed powerful from within and could become a great agent of change once they realised their potential as a significant change-making force. Singh said,

> We knew for sure that initially, we needed only one or two women to come forward and show courage. Once that happens, others too will follow. Besides, we had a good team of female staff members and volunteers who knew their languages coming from the local settings and therefore, were able to get access to their households without major difficulties.

Recollecting the days when they were pushing hard to motivate the women to come forward and unite for the cause of water, Singh said that they thought of creating groups at village level which would be called *Pani Panchayat* (which would translate into "water council" or "water collective" in English). Elaborating on the idea further, he said,

> People in rural areas understand the term *Panchayat* as it is used in *Gram Panchayat,* which is the village-level governing body to which people are elected through voting. The word *Panchayat*, thus, signifies some sort of authority and respect. Generally, representatives of

> the *Panchayats* would be men, and even if women are elected to the *Panchayats*, as *Gram Pradhan* (village head) it would actually be their husbands locally known as *Pradhan Pati* (husband of the village head) who would perform the duties on their behalf. No elected woman *Pradhan* would go to the block offices where they would get their funds for carrying out small development works under various government schemes.[7] Even after being elected, they would remain inside their houses and let their husbands do their work on their behalf as it was about power which they think only men have the right to.

It was quite evident as when the Parmarth volunteers would introduce him to many *Gram Pradhans* in Uttar Pradesh villages and *Sarpanchs* as they are known in the villages of Madhya Pradesh, they would actually be working on behalf of their wives. Even when we (the researchers) wanted to meet some of such female Sarpanchs in the villages of Chhatarpur districts of Madhya Pradesh with the staff and volunteers of Parmarth, they would simply deny meeting saying "*Sarpanch Ji ghar pe nahihain*" (*Sarpanch* is not at home), whereas they are the actual elected *Sarpanchs*.

The Parmarth's project coordinator said that after a lot of deliberations at the office among the staff members and the volunteers, they decided to form groups at the village level and name them as *Pani Panchayat* with predominantly women as the members. This way the women would have a feeling that they also are a part of an important *Panchayat* wherein they can meet like the men in their communities do and discuss their issues – essentially those related to water – and can work towards finding out solutions in their own ways. "We had noticed that a major part of the women's life in the countryside of the region is spent in arranging water for the household," he said. Even as women are shouldering the important responsibility of procuring water, which was required for most of the household chores, they had absolutely no role in the decision-making processes in the villages. That's what, he said, gave them a sense of opportunity to reach out to the women, identifying those with the leadership qualities to spearhead the campaign for water management by mobilising other women. He said that having understood the value of water owing to their everyday struggle for it, the women were always going to be better managers and decision-makers when it came to valuing water and conserving it. Singh went on to narrate the story of how their persistent efforts yielded result,

> After regularly meeting the women for 4–5 times and telling them that if they come forward, the issue of water scarcity could be dealt to a great extent, some women expressed their curiosity to understand what kind of work they will have to do. We told them that with voluntary services we would work for the rejuvenation of the old water bodies and create new ones to stop rainwater run-off during the monsoon and improve the groundwater level that will minimise their efforts for procuring water

> and save them a lot of time and energy. The idea clicked with some of the women. And the campaign first began from the Moto village of Lalitpur where 7–8 women decided to dig an old well which was completely covered up with soil. They kept digging despite people making mockery of them. Later, when they did not give in, other women and men too joined them after around 20 days. And finally, they were able to strike water after digging for around 100 days. Following this, they went on to create two check dams in the village to stop the rainwater run-off that resulted in increasing the groundwater level, which saw the water level in well improve. Similarly, stories of women leading such initiatives started to emerge from other villages too, and the campaign took off.

After the formation of the *Pani Panchayat* groups at the village level and building the cadres of the *Jal Sahelis* to lead those groups, Parmarth officials realised that the work was only half done and for ensuring the availability of water in the villages, it was crucial to bring the other stakeholders together. So, they started leading the *Pani Panchayat* members and *Jal Sahelis* to the district administration offices, block development offices, and *Panchayat* representatives so that they could be apprised of the gravity of the problems. They would call such activities as "social action." The regular meetings with officials and local public representatives helped Parmarth and *Pani Panchayat* groups to sensitise the officials at the district and block administration offices as well as the *Pradhans* at the Panchayat levels. With the passage of time, the authorities started extending administrative support to their work of water management and providing benefits under various schemes which the villagers were entitled to, like employment under Mahatma Gandhi National Rural Employment Guarantee Act (MNREGA), pensions, etc. Many water conservation structures were created and renovated through MNREGA funds.

In many programmes that Parmarth would organise for training and capacity building of the *Pani Panchayat* members and the *Jal Sahelis*, they would make sure that government officials like Collector, Sub-Divisional Magistrate (SDM), Block Development Officer (BDO), and others came and participated. This would help Parmarth bridge the gap between the administration and the rural population. Such events would also give women the confidence to interact and talk to the government officials with ease which was otherwise entirely in the realms of impossibility. And with the women starting to take the leadership roles in the initiatives of water conservation and management, the change was beginning to take place.

Stimulating Change through Participation, Engagement, and Empowerment

The coming together of the women with the tireless efforts of the Parmarth's project coordinator, cluster coordinator, team leader, and volunteers were expected to bring about some change in the status quo but "such an unprecedented change as it was seen in the Moto village was something that even we did not imagine," said Sanjay Singh, secretary, Parmarth. Imarti Devi of the village who was one of the three women to start digging the well initially told during an FGD that initially after the meetings with the *Bhai Sahab* and *Didi* (which she meant for programme coordinator Manvendra Singh and team leader Vineeta Yadav), there were 8–9 women who agreed to begin the work of digging the well. Narrating the story further, Imarti Devi said that out of 7–8 women, only three continued with the work as others backed out after being reprimanded by their husbands and family members. Among the other two women were Phulwati and another one who, as Imarti Devi informed, went to Ahmedabad with her husband who worked there after migrating from the village.

As Imarti Devi went on to share those experiences from 2012, her language changed to Bundeli from Hindi (which she was consciously trying to speak as she knew I was from outside). "*Bhai Sahab thakur saban ne jhagda macha dei haman ko ghar ma. Ken lage lugaai kuan khode aur tum...*" (Brother, all the Thakurs caused tensions in our homes. They said to our husbands that your wives have started digging well and you ...). "And then the men started talking among themselves that if women took the leadership roles then what will happen to them," she continued. On being asked, who said this, she said, "our husbands." Phulwati immediately cut her and said, "Not just our husbands, all the husbands in the village. It became an issue of concern for all the men in the community." As they persisted with their efforts for accomplishing the task, many women, children, and men joined them and withdrew as the days passed by without hope of striking water any sooner. It took them almost a hundred days when they started sensing the availability of water as the moist soil started to appear.

I asked them if they did not feel disheartened at any point of time during this long period of 3–4 months. Imarti Devi said, "*Bhai Sahab, haman ne sochoki agar haman ne karbe bhaye toh je agar jo nai kar paayein to apan ki beizzati huija*" (Brother, we thought that if we have started the work and could not finish it, it will be our insult). They said that what encouraged them to keep their efforts on was the fact that people had started extending their support. Even if someone left out for some reasons, some others joined. Initially, they said that after a few days, some women started sending their children for their help who would help the women in disposing of the dug-up soil and its disposal. Then some older men started extending their helping hand, which they said was a significant boost to their confidence. On being asked for how many hours a day they worked digging the well, they informed every day they started the work at 8 a.m. and worked till 5–6 pm and this went on for four months when the well was completed. After two months or so, all those women who had withdrawn initially joined hands as they realised that they would not stop until the well is dug. Later, with financial support from Parmarth, cement, sand, and rocks were purchased for the completion of the well.

After the construction of the well was complete, in order to make sure that the water level was maintained in it, they immediately started the work of constructing the check dam so that rainwater could penetrate under the soil. Many men and women came forward and offered voluntary labour work for the construction, along with some of the hired workers and labourers. The resources like construction materials were provided by Parmarth, and in the same year, they erected the check dam. This not only saw water level of the well surging but the hand pumps in the village which had run dry started supplying water. After one check dam, they constructed a second check dam with the view to hold and conserve as much rainwater as possible so that the village could have water in the wells and the hand pumps throughout the year. These efforts did yield results as they did not only get water for household work through wells and hand pumps but also had adequate water to irrigate their agricultural fields.

The stories of struggles of women for water were no different in Kakradi village of Lalitpur who would travel for three kilometres for water. After they were mobilised for the formation of *Pani Panchayat*, they started knocking the doors of the authorities from the local administration. Initially, the authorities would hardly pay heed to their concerns, and they would be made to wait for hours outside the offices of the district collector, SDM, and others. Tired of such treatment, the members of the *Pani Panchayat* and *Jal Sahelis* including Rani, Kalan Devi, Sapna, and others went directly to Lucknow in the public hearing conducted by the chief minister (CM). In the meeting, they registered their grievance to the then CM Akhilesh Yadav who immediately took action and directed the district collector of the Lalitpur district to address their concerns. Following this, Rani, a *Jal Saheli* of the village, said that the pond in the village which had dried up was deepened, cleaned, and widened. The channels through which the water would flow to the pond during the monsoon were cleared too. As a result, in the monsoon of 2012, the pond got enough water that lasted for the entire year even as the rain was deficient that year. Rani informed, "Initially when didi came to our village we were reluctant to form *Pani Panchayat* here. But when they told us the story of the Moto village, we thought we too could come together for making our lives better."

Vineeta Yadav of Parmarth said that the women of the village whom they first met were little educated, and therefore, they understood things better and quicker. They themselves chose the other women who could work together as a team and got united. In the beginning, they were around ten women, and the number grew, and today they are 30 in the village. Kalan Devi, a *Pani Panchayat* member of the village, informed that after the deepening of the pond, they also did large-scale plantation around the pond, and started a campaign for keeping the village clean by cleaning the streets, nullahs (drains), etc. Rani's energy level was worth noticing when she informed that earlier the village did not have toilets in the households, but after the interventions of the *Pani Panchayat* with the community members in the village and with the local administration, today, 75 per cent of the houses have toilets constructed, and many are getting them constructed.

Pani Panchayat member of the village, Kalan Devi who is in her late 50s says,

> We are old, so we do not travel to distant places like other members of the group, but women like me extend our whole-hearted support to the work being carried out by *Pani Panchayat* in the village. Sometimes, when the sanitation workers don't come to the village for several days, we ourselves hold a meeting and then start cleaning the streets of the village to ensure our village look clean.

When asked what kind of change she has seen in the village after the formation of the *Pani Panchayat*, Kalan Devi said, "*Bahut badlaav hue hain. Sabse badi badlaav toh jei hai ki humein gaon mein paani milne lage. Pahle humein dhai-teen kilometre paidal chalne padte the paani ke liye.*" (Many changes have happened. The biggest change is that we have got water in the village itself. Earlier, we had to walk on foot for 2.5–3 kilometres for water.) In addition to mentioning the work like toilet construction, sanitation work, and kitchen garden, she also said that the group is now creating a fund with the contribution from the members and are planning to help some of the women in the villages who have some livelihood crisis to purchase cattle like cow and buffalo so that they could produce milk, sell it, and earn money. She said that *Jal Sahelis* have talked to the local administration about this and they have also assured for some monetary help through the bank. So, with the support of the administration and the money collected by the group, they would be able to help such women and families in the village.

The members of the *Pani Panchayat* including Rani, Kalan Devi, Sapna, and others informed that they organise a meeting of the group on the 15th of every month to discuss the situation of the village. They informed that they were also trained on how to repair a hand pump and how to write an application, get signatures of the villagers on it, and submit it to the authorities. Rani said that for small problems they themselves repair the hand pumps, and if there are some major technical problems, then they call the mechanic and get them fixed.

Responding to the question as to what kind of changes have they experienced in their own lives after joining the *Pani Panchayat*, they said that earlier, they would not even move out of their houses as they thought that being women their duty was to stay at home and take care of the household chores and look after children. She said,

> After the initial meetings, we realised that women could also take part in some important work and contribute to bringing about change. As our interactions increased in frequency with the officials of Parmarth through meetings and training programmes, we started realising our

> potentials. We got rid of our inhibition of moving out and started going to different places to learn new things and improve our knowledge after looking at the work being done by women in different places. Once, we went to Jhansi, Lucknow and once we even went to Delhi in a protest to raise demands related to agriculture and water.

It needed a great level of sensitisation for men of the communities as well. Walking down memory lane, Rani narrated how her husband would stop her for not going outside for such meetings as it was against the reputation of the family and the social traditions. She said,

> Didi and Bhai Sahab (female and male representatives of Parmarth) would convince the men of the village to allow their wives to go out saying if the women get empowered, the generations to come will benefit. They would give them examples of the women and daughter of other villages who were involved in various kinds of work for social welfare. It took some time for them to understand the benefits of women moving out, forming the group and organising meetings. Today, they do not object when we do our work; rather, the men have started supporting us a great deal in our work.

And similar stories emerge from different villages of the region including Sopta, Udotpura, and Itaha Kalpi; villages of Jalaun, Kalothra, and Vijrautha; and other villages of Lalitpur where Parmarth has built strong cadres and networks of *Jal Sahelis* and *Pani Panchayat* with the help of its grassroots level workers including the cluster coordinators and team leaders. In all these villages, the women have not only been able to address their water woes to a great extent in various ways but have also mobilised and sensitised the communities over the issues of sanitation, health, education, etc. In many such villages where *Pani Panchayat* groups are working actively, women have started sending their girl children to schools. They have been able to secure pensions for the old-age people and the widows of their villages, employment for themselves under the Mahatma Gandhi National Rural Employment Guarantee Act (MNREGA), houses for the people below poverty line under Pradhan Mantri Awas Yojana – Gramin (PMAY-G), and others. Parmarth keeps making efforts to keep the *Pani Panchayat* members and the *Jal Sahelis* motivated so that they could ensure the sustenance of the changes that they have brought together.

Sustaining the Change: *Jal Sahelis* Taking the Work Forward

Almost eight years have passed since the project "Establishing Women's First Right on Water" was started by Parmarth in 96 villages of three districts of Bundelkhand. Many villages have been transformed with the constant efforts of the Parmarth workers at different levels, i.e. from the state coordinator to the volunteers at the cluster and village level for mobilising communities, especially women, and encouraging them to take action. There are innumerable villages in the region where women's struggle for procuring water in their everyday lives is there for everyone to see. The women in such villages have accepted those situations as realities of their lives without making any efforts to deal with them.

For Parmarth, as its secretary Sanjay Singh says,

> It took huge efforts for the organisation to see the change that it envisioned around ten years ago before the project was started. And if I tell you frankly and honestly, we have been able to achieve only 50 per cent of what we intended during the beginning of the programme because of several reasons like – limited resources, limited human resource, inadequate support from the government and administration, geographical features, climate change etc.

However, a sense of contentment could be seen on Singh's face when he said, "Nevertheless, with our limited capacities, whatever we have achieved has given us not only satisfaction but also an inspiration to scale-up such initiatives. And we know that now, we also have to ensure that whatever change has happened with our efforts needs to be sustained." On being asked what do they do to ensure sustainability of the change that they have brought about he said,

> There are many villages in the region where several *Jal Sahelis* have taken the responsibilities of keeping the *Pani Panchayat* groups intact and carry forward the good work. In those villages, we believe our interventions are least required. They themselves reach out to our district offices when they need our involvement for any task that they want to accomplish. They regularly do a consultation with our cluster and district level officials.

He gave examples of such villages as Kalothra, Kakradi, and Moto. For those villages, where members of *Pani Panchayat* and *Jal Sahelis* are not as active, they hold a monthly meeting to make sure that these women meet with the women of the other villages where they are more productive and carrying out inspiring work. Besides, the representatives of Parmarth also visit those villages at least once every month to make sure their communication channels are active, and there are no communication gaps. And this practice, he said, also helped them in informally evaluating the works that were being carried out by the women groups in the villages and their efforts to maintain those which were carried out in association with Parmarth earlier.

The secretary said that they consider the work of developing *Jal Saheli* cadres and forming *Pani Panchayat* groups as the flagship programme of Parmarth for which they were recognised not only nationally but also internationally. Representatives from Parmarth are invited to different forums on the national and international level to share the success story of the project. "Many researchers and students came from different countries to study our work, meet the *Jal Sahelis* and understand their work," Singh said. Therefore, we are making all kinds of efforts to make sure that the good work carried out by these women continues in the days to come.

Conclusion

Kumar, the UP state coordinator, said that in an extremely backward region like Bundelkhand, problems are much more complex than they appear superficially.

> Through our experiences of working in the region for more than two decades, we have learnt that for any initiative to be successful we need to understand the problem as well as its social context and background. It is only then that we could develop a good programme that could address the complexities embedded in the nature of a problem.

Citing the example of the project of "Establishing Women's First Right to Water," he said that initially, they tried to bring men and women together in the meetings and hold discussions so that both could be sensitised about the issues of water and the hardships that women faced for collecting water from distant places. But after several deliberations, they found that women were very reluctant and shy in opening their mouths in group meetings when men were around. "That was when we thought of forming a separate group of women where they could come together speak among themselves and learn to articulate their issues," Kumar said adding, "We thought this might help them gain the confidence to assert their voice before anyone. And it worked. We were surprised to see the kind of transformation that was happening among the women and the kind of agency that was developing in their groups." Today, many of the *Jal Sahelis* can be seen talking about their issues even to the district magistrate (DM) who is the highest authority at the district level.

Vineeta Yadav, the team leader, said that uniting the women was the most important thing so that they could go to the local authorities in large numbers. Until they come in large numbers, the officials would not take them seriously as they are in the habit of dealing with the people's issues in as casual a manner as one can imagine. The best they could do to address one's problem is to ask them to give a written application and then keep them in a file which will remain unattended forever.

Yadav said,

> So, we convinced the women to expand their groups and ensure that maximum women came together whenever they had to visit a local

> administration office. Having worked with the women of Bundelkhand in other areas for different initiatives, we knew it well that after meeting with the women and talking to them for a few times they would realise the need to come together and take required actions to see the change happen. Over a period of time, we have realised that when women decide to accomplish a particular task, they do that with a lot of commitment and persistence. We have worked with women on the issues of water even before this project.

Recalling one particular incident, she said,

> Once, we were organising a *dharna* (sit-in) at the DM's office for the water crisis in Itaha Kalpi village of Jalaun. The women also took their empty earthen pots with them. When the DM did not pay any attention to their protest for a few hours, they broke their pots before the DM's office. The DM, then, immediately listened to their problem and ordered the officials to address their concerns. After that, the wells in the village were cleaned as the water in them was extremely dirty, and the re-boring of the hand-pumps was also carried out.

These works of *Pani Panchayat* women and *Jal Sahelis* have been stories of their never-say-die attitude and perseverance. They had a difficult path to tread with people from their own community, trying to pull them back when they decided to get into action for addressing their water-related hardships. While the women of the Moto village wanted to dig an old well on their own, the men from their own village would taunt them and try to demoralise them saying digging of well is beyond the capacities of women. Similarly, when the *Pani Panchayat* members of the Kalothra village demanded their share of work under MNREGA from the local communities, they deliberately gave them difficult tasks like removing boulders, breaking them, etc. so that the women would themselves give up. These women said they were well aware of the intentions of the authorities behind giving them such heavy tasks and, therefore, they had resolved not to succumb under any circumstance. They said that even though the work was tiring and exhausting, they made sure that they finished them without any complaint to the administration as they wanted to prove that women were not inferior to men in any aspect of life. Seeing the willpower of the women, the authorities too realised their mistake and appreciated their efforts in addition to giving them regular work under the employment scheme.

These stories indicate that Parmarth and its officials with their experiences of working with the women of the region for more than two decades had been able to successfully mobilise women with a sound understanding of the socio-economic conditions and cultural settings which help them design effective interventions to work on issues related to water and the associated struggles of women. Although Parmarth uses the concept

"Theory of Change" only while developing project proposals seeking grants from donors, the way its officials and workers engage with the communities in a given geographical region for addressing their problems does reflect a theory of change approach. They start their work by assessing the existing situations comprising the culture, norms, behaviour, attitude, and awareness among people about a given issue. After the assessment of the situation, they engage with the different relevant stakeholders ranging from the community members to the local public representatives, civil society actors, activists, and the government authorities at the block and district levels which leads to the creation of "pre-conditions" which are, as James (2011) points out, significant for bringing a long-term desired change. These pre-conditions include coming together of all the stakeholders and regular interactions and dialogues among them for deciding the suitable course of action for the change that they intend to bring about. These dialogues and deliberations facilitate an atmosphere for open conversation among the village community members and also with the local public representatives at the Gram Panchayat level and officials of the local administration who have a role to play in the desired plan of action. Such interactions, as James (2011) observes, improve the partnership among the stakeholders and also empower them to be more active and engaged in the change initiatives.

Isabel Vogel (2012), underlining the theory of change approach for an effective intervention, says that for any change intervention, an organisation has to understand the context appropriately, identify the stakeholders, and then carve out a chain of events linked to each other in a logical way that could see the long-term change occurring. This approach clearly reflects in the design of the interventions of Parmarth for addressing the water-related issues in many villages of the Bundelkhand region as they study the context of the problems as well as ensure the participation of the actors who could influence the behaviour and attitudes among the communities for the desired change. Leat (2005) also says that the theory of change should begin by articulating a theory of problem, find out the causal story, and then include the defining actors who are capable of influencing the story of change which is visible in the interventions of Parmarth.

Singh, the project coordinator, said, "Even as there are government schemes to develop water conservation and management infrastructure, the way they are done with the nexus of the government officials and the contractors hardly makes any difference." He said that when the government gave work like water-shed development to the contractors in Chhatarpur and Tikamgarh districts of Madhya Pradesh, none of them addressed the issue of water scarcity as they totally ignored the local knowledge of the communities.

When the higher authorities realised this, they decided to involve local NGOs and work in their partnership. This way, when Parmarth got the work of water-shed development in Chhatarpur, the focus of the work shifted from a one-way top-down approach to a network perspective involving all the relevant partners. The work carried out by Parmarth for water-shed development in Chhatarpur, which will be discussed in detail in the next chapter, helped a great deal in addressing the issue of water crisis in select areas. These interventions appeared to be in line with the argument of Rick Davies (2004) wherein he calls for removing the one-dimensional nature of change processes and creating a network of events through a network of actors for developmental projects. Having realised this aspect long ago, Parmarth ensures the participation of all relevant actors in its interventions and makes all efforts to make the local communities the most prominent actor who could not only contribute during the implementation of a programme but could also take care of the assets developed during the intervention for their longevity and sustenance as they are not only the beneficiaries but also the change-makers.

Notes

1 Also known as global goals, SDGs were adopted by UN member states in 2015 as a universal call for action to end poverty, protect the planet, and to ensure peace and prosperity for all by 2030.

2 The Notebooks of Leonardo da Vinci, translated by Jean Paul Richter (1888).

3 Rick Davies, April 2012: Blog post on the criteria for assessing the evaluability of a theory of change. http://mandenews.blogspot.co.uk/2012/04/criteria-for-assessing-evaluability-of.html.

4 ESPA guide to working with Theory of Change for research projects produced by Isabel Vogel in association with LTS International and ITAD or ESPA Programme.

5 Developed by John Turner and his colleagues, self-categorisation theory seeks to interpret and describe how people identify themselves with a social group based on the underlying attributes that they share which forms their emotions, behaviours, and attitudes.

6 Study on Women's right to water: Capturing the Realities on the Ground in Three Project Districts of Jalaun, Hamirpur and Lalitpur (2011) produced by WFS in partnership with Parmarth and supported by EU.

7 In Uttar Pradesh one-third of the total Gram Panchayat seats are reserved for women.

4 Communication for Holistic and Sustainable Water Governance

Forging a Green Public Sphere

Water is directly linked to the issues of livelihood in the rural areas of India where a high percentage of population is dependent on agriculture and livestock production that requires water. The lack of water simply means a lack of ways of securing basic necessities of life like food, clothes, and shelter (FAO, 2008). It was the supply of water provided by the government in India during the 1960s and 1970s that brought the green revolution in India and helped millions of poor farmers improve their lives and livelihood besides alleviating poverty to a great extent (Hussain, 2007). Today, water has once again become a scarce natural resource in many parts of India, causing livelihood crisis and forcing millions of people to migrate to different places in search of work. Shortage of water and other environmental issues have become a serious concern of our times that need immediate human attention and action. And headway in this direction could only be made through education, communication, and public awareness which can help in redressing the existing trends and practices that are threatening the future of humanity on earth (UNESCO, 1997).

Scholars world over, today, are talking about the need for crafting telling stories related to environmental degradation for human society to act fast. However, they believe that because of the complexities involved in the stories and the resulting depression among the readers and audiences, the stories related to the environment and climate change do not sell and fail to bring good ratings to the television news channels (Crist, 2019). According to Crist (2019), concerned activists, scientists, and journalists across the globe feel that even as the issues related to environment and climate change are emergencies, there is no one paying heed to the stories related to them. For those who have not been a victim of climate-related catastrophe, the issue of climate change and environmental degradation seems to be irrelevant, he feels.

However, Anders Hansen (2015) believes that media and communication play a vital role in constructing environmental problems as the issues of political and public concern and action. He says that media have the capability to push the environmental processes to get recognised as the issues for the larger society to think about. For Hansen (2015), environmental issues

DOI: 10.4324/9781003350286-4

do not manifest on their own, but it is through the claims made by relevant individuals and institutions that they get the due recognition in society. This shows the significant role that communication must play in bringing environmental concerns at the centre of the public discourse.

The news media have been finding it challenging to incorporate stories related to the environment and climate change – that could be instrumental in influencing social practices, public opinions, and understanding about the issue – because of several constraints like shrinking news spaces, the political economy of the news organisations and changing ownership patterns and other competing issues (Friedman, 2004; Boykoff & Roberts, 2007). However, there have been initiatives by civil society groups, NGOs, and concerned individuals which have helped the cause to a certain extent at least in some select areas. They have not only helped in making people aware of the gravity of the situation but also mobilised them to understand the urgency and act in their own ways to address the visible issues in their local settings.

This chapter discusses such initiatives in pockets of Bundelkhand where Parmarth is working on the issues of water and livelihood with the support of some donor agencies, local administration, and communities. The first part of the chapter discusses environmental communication as a discipline in the field of communication and would underline the relevant scholarship in the field that informs this study. The next section of the chapter would capture the communicative activities of Parmarth for water-related interventions. It then enumerates the mobilisation and communication activities being carried out by Parmarth around water. The following section brings to the fore how women play a role in the process of mobilisation and communication. The next section of the chapter is on the WADI project which supports tribal communities in developing fruit orchards for livelihood; and promotion of kitchen garden practices among the local communities. One of the philosophies behind the project is also to ensure green recovery of the nature through plantation, water resource development, and soil conservation. And the final section of the chapter is the conclusion. In doing so, the chapter would use environmental communication as a theoretical framework to interpret, analyse, and explain the interventions of Parmarth as their work is both directly and indirectly related to environment, ecology, climate, sanitation, livelihood, etc. As argued by Cox (2007), as a discipline, it is the ethical duty of environmental communication to enable society to "respond appropriately to environmental signals relevant to the well being of both human communities and natural biological system" (p. 5).

Environmental Communication: Understanding and Defining the Field

Humans have been communicating about the natural world, and interacting with it since civilisations started to evolve on earth. Rachel Carson (1962) in her book *Silent Spring* says that since the beginning of life on earth, physical forms and habits of plants and other species have mainly been moulded by the natural surroundings and not the other way around. However, in the last century or so, the effect has been the opposite, as humans have become significantly powerful in terms of altering the natural environment that they inhabit. While expressing her deep concerns for the deteriorating environmental conditions, Carson brings to the fore the excessive human interventions in the natural world that caused the contamination of life-supporting systems like air, earth, rivers, and sea with dangerous and lethal materials. Her work is considered seminal in terms of triggering concern and debates about environmental issues. Abbati (2019) feels that Carson's literary production had a lot to contribute to the birth of environmental communication as an academic discipline. For him, the "green impulse" provided by Carson resulted in the research and analysis of environmental problems and the search of ways to address them in the 1960s, 1970s, and 1980s till the present time also led to the formation of the International Environmental Communication Association (IECA). Studies suggest that environmental communication emerged as a field of academic study in the last few decades as a result of the pressing ecological problems, primarily caused due to human interventions in nature (Anderson, 2015). Many scholars trace the beginning of the term "environmental communication" in academia to the early 1980s in the United States. The term is thought to have emerged following the analysis in a 1984 publication by Christine Oravec, of the controversies in the discourses in the early 1900s over the construction of a dam on a highly regarded natural site (Cox & Depoe, 2015; Milstein, 2009; Cox, 2012).

Milstein (2009) defines environmental communication as "a field within communication discipline, as well as a meta-field that cuts across disciplines" within which the research and theory are "united by the topical focus on communication and human relations with the environment" (p. 344). While in general, coverage of environment-related issues in the media is also considered a form of environmental communication, the term has a tradition of being identified as institutional communication of a state about environmental policy (Nofri, 2011). Several scholars contest the use of the term "environmental communication" for the coverage of environment in media and consider it to be mere "environmental information." Others prefer using the former, arguing that when "information" is conveyed with the aim of letting someone know about something it is communication (Nofri, 2011).

Robert Cox (2012) finds it difficult to define environmental communication for the wide variety of subjects that it deals with. Cox calls for a need to focus on "symbolic action" to offer a richer definition, and he uses the term *environmental communication* to refer to "the pragmatic and constitutive vehicle for our understanding of the environment as well as our relationships to the natural world; it is the symbolic medium that we use in constructing environmental problems and negotiating society's different responses to them" (Cox, 2012, p. 20). Pezzulo and Cox (2018) believe that environmental communication could not be defined merely as information and talking about the issues pertaining to the environment but an appropriate definition of the subject must take into consideration the "roles of language, visual images, protests, music, or even scientific reports as different forms of symbolic action" (p. 34). For Pezzulo and Cox (2018), information is not the only important aspect of communication; and therefore, they describe the term environmental communication as expression performing pragmatic and constitutive functions as Cox (2012) pointed out. Environmental communication is pragmatic because "it consists of verbal and non-verbal modes of interaction that convey instrumental purpose. Pragmatic communication greets, informs, demands, promises, requests, educates, alerts, persuades, rejects, and more," and it's constitutive as "it entails verbal and non-verbal modes of interaction that shape, orient, and negotiate meaning, values, and relationships. Constitutive communication invites a particular perspective, evokes certain beliefs and feelings (and not others), fosters particular ways of relating to others, and thus creates palpable feelings that may move us" (Pezzulo & Cox, 2018, pp. 34–35). Corbett (2006) too calls for the need to define environmental communication beyond the exchange of words and pictures about the environment as our actions too constitute a form of environmental communication in some way or the other. She considers it as a "complex and multi-layered phenomenon" and says that messages about the environment have "ideological roots that are deep and that are influenced by individual experience, geography, history, and culture" (p. 23).

Maitreyee Mishra (2013) traces the environmental communication in the folklore, through which the indigenous groups, that include the Native Americans, the Africans, the Indians, and the Chinese, communicated with their people about the earth. Environmental communication is said to be in existence since the time human beings started interacting with each other and with nature. However, its beginning as a special discipline could be traced to the 1960s when communication practitioners and academicians gave it a label (Jurin et al., 2010). In their book, Jurin et al. (2010) mention the beginning of the publication of *Journal of Environmental Education* as a significant development in the field of environmental communication. They cite Schoenfeld (1969) who defined environmental education as "communication aimed at producing a citizenry that is knowledgeable concerning our environment and its associated problems, aware of how to help solve those problems, and motivated to work toward their solution." At the start of the journal around four decades ago, environmental education and communication were considered as conjoined being the two facets of the same coin (Jurin et al., 2010).

In many developing countries, there is an emergence of environmental or green public sphere owing to the rising environmental discourses or "greenspeak" (Yang & Calhoun, 2007). These discussions engage people in debating environmental concerns as well as the dangers of irresponsible behaviour of human beings towards nature. A similar observation is made by Cox (2012, p. 25) when he says that whenever "we engage others in conversations, questioning, or debates we translate our private concerns into public matters and thus create spheres of influence that affect how we and others view the environment and our relation to it." He adds that such translations of our private concerns into public matters in different forums give rise to an "environmental public sphere" (p. 25).

In the last few decades, environmental problems have outgrown the systems and infrastructure put in place to solve them (Najam et al., 2007). As a result, environmental communication scholars were faced with a challenge to not only critique the existing discourses in the field but also to facilitate public processes for producing alternative discourses that were more relevant and sustainable (Milstein, 2009). Gradually, the discipline adopted and brought into application various theories like theory of media construction of reality, framing analysis, social movement theory, cultural theory, rhetorical theory, and other theories of media and communication, discourse analysis, etc. to study different aspects of the subject. Environmental communication is vital for humanity which seems to have shunned the nature in its continuous quest for growth and development as it not only explains about the ecological damages occurring due to excessive human interventions in the nature (Milstein, 2009) but also helps in shaping our understanding of the nature and our relationship with it (Mishra, 2013; Cox, 2012). The discipline is also significant in terms of addressing violent conflicts originating out of the environmental crisis and working for ensuring good health of the environment by discerning and promoting good practices in the field (Meisner, 2015).

Among the major thrust areas within the domain of environmental communication, shrinking natural resources like water, forest, fertile agricultural lands, etc. find an important place. This chapter would attempt to explain how Parmarth makes use of communication as a tool to mobilise communities around water-related problems and motivate them for collective action to resolve them.

Communicating about Water: Initiatives for Addressing Water Crisis by Parmarth

For any organisation wanting to engage in some change-making process, understanding the communication process and its various facets is imperative. Without having a sound understanding of the complex communication processes and the underlying challenges, it would be extremely tough to bring about the desired change that it is seeking to accomplish. To make people in a given locality, partner for any environmental conservation effort, the foremost idea should be to realise how they make sense of the natural world that they live in so that one could make effective messaging and communication strategies to make them a party to the change-making process and for sustaining the change (Morgan, 2003).

For effective implementation of its water-related programmes, Parmarth has formed community-based organisations (CBOs) at community levels in the villages. Such organisations are given different names in different areas like *Pani Panchayat* (water council), Gram Chetna Samiti (village sensitisation committee), Village Water-shed Committee, etc. Parmarth, through regular communication and interaction with the members of the CBOs, which is mostly interpersonal or group communication and dialogic in nature, tries to understand the ground realities. It gains insights through the experiences of the CBO members and their narrations and then with their inputs and wisdom combined with its own expertise and resources mobilised through various sources designs interventions for addressing the water-related issues. Through formal and informal meetings, training programmes, seminars, and workshops, it aims at building their capacities for their effective involvement in water management programmes. These CBOs are then responsible for conducting dialogues with the other community members in the village to mobilise them for their support and contribution. Many times, the organisation's officials also directly go to the villages and hold meetings and dialogues with the community members through the CBOs.

Mobilisation and Communication Activities

To understand and address the issue of drought and water crisis, the organisations have to study, plan, and then make strategies to effectively deal with the complex process of educating, mobilising, and making the communities act to implement a water management project in a particular area. Parmarth too has to first invest in reaching out to different villages and identifying the most vulnerable ones where the interventions are immediately needed. They, then, chalk out the communication activities to mobilise the local communities to seek their involvement for the longevity of the water conservation structures being constructed. Such activities carried out by Parmarth include rallies, *yatras*, formal and informal meetings, trainings, workshops, seminars, symposiums, cultural programmes, special programmes on special days, etc. Different activities are selected for different occasions. What follows is an analysis of how these activities help the NGO in communicating effectively with the people.

Yatras: *Yatras* (processions) are one of the common activities that Parmarth organises to bring awareness among the people and to take up their concerns with various stakeholders like government authorities and media for redressal. The district coordinator of Parmarth for Chhatarpur district of Madhya Pradesh, Ajay Yadav, informed that these processions are also of different kinds depending upon the objectives with which they are organised. "I myself have been a part of many kinds of *yatras* in different parts of Bundelkhand," he said. On the basis of information given by him about the *yatras*, some of them could be distinguished as below:

- **Jal Yatra:** *Jal Yatras* are carried out by Parmarth workers and local youths to educate and sensitise people in the villages to save water and work collectively to save water bodies to achieve prosperity. The youth are also included from the local music bands who keep singing and playing music during the *jalyatra*. The participants prepare music and songs to creatively convey their messages for water conservation and at times they also prepare, on the spot, songs and music according to the existing water-related problems in the villages, and play them. The last *jalyatra* was carried out by Parmarth from September 14 to 18, 2016, covering 20 villages. The participants carried musical instruments like harmonium, *dholak*, *jheeka*, and *manzheera*. It had one singer and choir to give their messages through singing. They sing songs of different folk genres mostly in bundeli dialect like *Pani hai apni zindagani, pani khan bachane* (water is our life, let's save water) – which is from a rural folk genre called Rai, and *Medhan paudh lagane, jeevan khushhaal banane* (we should plant trees on farm bunds and make life prosperous) – which is from a folk genre called Badhayi. While the songs of Rai genre have fast and loud music, Badhayi is a genre which is slow in rhythm and songs are mostly emotional. Yadav said that the different genres have different significance. Rai songs are sung to attract people. When people hear the loud and fast music it fills them with energy and thus, they come and join the *yatra* in villages while Badhayi songs have an emotional appeal in them that touches the hearts and minds of the people on the issues raised in the songs, he says. On being asked why they choose to conduct *yatras*, he says,

  > The vehicle bearing the posters and banners from all sides, music and the sound system used in the *yatra* help us pull crowd in villages at one place where we could talk and listen to them. We also distribute printed materials to the villagers like pamphlets and leaflets for spreading awareness towards protecting water resources and give them a form to fill enlisting the problems they face in the villages.

- **Jal Hal Yatra:** Parmarth also networks with other organisations working in water and environment sectors to deal with serious issues. Parmarth organised a *Jal Hal Yatra* (water solution processions) in association with organisations like Swaraj Abhiyan, Ekta Parishad, National Alliance of People's Movement (NAPM), Rashtriya Jal Biradari, and others. The 10-day *yatra* which was carried out in the

Marathwada region of Maharashtra and the Bundelkhand region of Madhya Pradesh and Uttar Pradesh aimed at spreading awareness among the people of the drought-hit regions about their entitlements during droughts as per the orders of the Supreme Court. The *yatra* saw the participation of noted activists and social workers like Yogendra Yadav, Medha Patkar, Dr. Sunilam, Rajendra Singh, PV Rajagopal, and others, and it caught the attention of the local authorities and representatives, said the Madhya Pradesh (MP) state coordinator of Parmarth, Rupak Ghosh. The *yatra* was carried out in around 36 villages of the Bundelkhand region from the Tikamgarh district in Madhya Pradesh to Mahoba in Uttar Pradesh during which *Nukkad Sabha* (street meetings) and *Ratri Choupal* (night meetings) were organised in different villages to understand the challenges of the local dwellers through dialogues and discussions and to plan activities to redress them.

- **Lok Samvaad Yatra:** Such *yatras* are organised especially before the elections so that the pressing concerns of Bundelkhand like water scarcity, drought, migration, farmers' distress, and livelihood issues could find mention in the political discourse at both the state and the national level. *Lok Samvaad Yatra* would translate into English as a public dialogue procession. Such *yatras* were organised in different parts of Uttar Pradesh as a part of the *Jal Jan Jodo Abhiyan* (water people alliance campaign) being undertaken together by Parmarth in collaboration with Tarun Bharat Sangh and Teesri Sarkar for conducting dialogue (*samvaad*) with the political parties in the light of 2017 assembly elections in the state and to press for the inclusion of environment- and water-related problems in their manifestoes of the political parties during the elections. Manish Kumar, the state coordinator of Parmarth for Uttar Pradesh, informed that the main aim of these *yatras* is to spread awareness among the voters about foregrounding their issues, and to also publicise the concerns through media. At the same time the effort is to pressurise the candidates fighting elections and the political parties fielding them to deliberate on how they are going to address the people's struggle for water, agriculture, health, etc. Such *yatras* were carried out extensively in various parts of the state, especially Bundelkhand, before the state assembly elections from December 18, 2016, to January 14, 2017. Similarly, even during the general elections of 2019, Parmarth organised such marches and conveyed to the political leaders contesting the elections that people would vote only when they are given assurance that their water-related sufferings would be addressed. In addition to that, it ran a widespread campaign among the people of the Bundelkhand by roping in the theatre arts students from the Bundelkhand University and National Service Scheme (NSS) members to educate voters on voting for the right candidates who have a vision to work for the development of the region. The students went from village to village to run the campaign and conducted activities like street plays, skits, singing, public meetings, etc. and also distributed printed materials to spread awareness.

Training Programmes and Workshops: Training programmes and workshops are another important way for organisations to establish communication with the stakeholders, especially the grassroots level workers, and to strengthen their capacities for effective implementation of the initiatives. Among such activities organised by Parmarth are the following:

- Workshops for social organisations and social workers are organised to train them to create pressure on candidates from different political parties contesting elections at all levels from Lok Sabha to the Panchayat elections time-to-time. Such workshops were organised by Parmarth during the 2017 assembly elections in Uttar Pradesh to include the issues of environmental degradation, water scarcity, drought, unemployment, etc. in their election manifestoes. Similarly, it was done during the Panchayat election in Madhya Pradesh in 2015 and General elections in 2019. Unlike *Lok Samvad Yatras* which happen on the go as they travel to different places as a procession, these programmes are organised at specific venues where participants are invited from various sections of society. They also invite local journalists to such workshops who share their insights about the existing problems so that they may keep reporting about them on a regular basis. Parmarth officials say that inviting the journalists not only helps them talk about the pressing issues effectively but the activities and discussions also get covered in the media, which helps their cause of taking the messages across to the political parties during the elections.
- Documentation workshops are organised at regular intervals by the organisation to train workers from various social organisations on effective documentation practices and record-keeping. The participants from different regions of Uttar Pradesh, Madhya Pradesh, Rajasthan, Chhattisgarh, and other neighbouring states are invited to benefit from these programmes. During the workshops, participants are explained about different aspects of documentation like preparing case studies, work plans, proposals, evaluation reports, etc.

Meetings: Meetings are the most common activities facilitating regular communication between the organisation and the other stakeholders that include community-based organisations (CBOs), community members, concerned government officials, other technical organisations, local committees, Panchayats, and more. Meetings are both formal and informal depending on the issues and the stakeholders. The state coordinator of Madhya Pradesh explained that the most effective way of communication in the rural areas of the Bundelkhand region is interpersonal communication or dialogic communication. He said,

> We have been able to address many issues only on the basis of interpersonal communication and dialogues that we hold with the community members at village levels. When we facilitate such dialogues between the community members and the officials from the local district and block administration, the authorities come to know about the challenges faced by the villagers and the seriousness of the issues. Local communities too feel confident and empowered when government officials come to their villages to talk to them directly. They freely communicate with them and sometimes also suggest the ways by which certain issues could be dealt with.

Most of the meetings which are organised regularly by Parmarth are meetings among the officials and its grassroots level workers; village-level meetings between Parmarth and the villagers; Parmarth and CBOs, local organisations, local authorities, and community members; *Pani Panchayat* meetings; water literacy camps; annual symposiums; etc.

Water Rallies: Rallies are also a commonly used mode of communication by Parmarth. They are considered effective when it comes to spreading awareness among the people and mobilising them for action. Thus, Parmarth organises rallies regularly in different parts of Bundelkhand for disseminating messages about using water judiciously, protecting water bodies, and guiding people in both rural and urban areas about their responsibilities to save the environment. Ghosh informed that in the Bundelkhand region of Madhya Pradesh, especially in the districts of Chhatarpur and Tikamgarh, 10–12 rallies are organised in a year on different occasions. Such occasions include World Water Day on March 22, World Environment Day on June 5, Women's Day on March 8, Gandhi Jayanti on October 2, and also on other occasions when there is a need to involve people in some event and mobilise them for action. Besides this, in connection with *Jal Jan Jodo Abhiyan* too, Parmarth organises water rallies in different parts of Bundelkhand to create awareness among people about the necessity for initiatives to save water bodies and to develop new structures for the conservation of rainwater in the region. If the district coordinator of the Parmarth for Chhatarpur district is to be believed, the rallies give them an opportunity to build rapport with the people in the villages who they seek to mobilise and work with. "When we take out rallies, people in the villages come and enquire about the purpose of the rallies, talk about the crises they face for arranging water and also about the ways in which these issues could be dealt with in association with all stakeholders," said Ajay Yadav. In the rallies, they involve local people, especially local youth, and students from schools and colleges who raise slogans according to the theme of the rally to spread their message. Rallies are followed by meetings with the villagers where the prevalent water availability scenario in the village is discussed and a preliminary roadmap to address the existing problems is worked out based on their extent and nature. Usually, while going to a particular village, the rallies will include locals so that people of that village relate to the message being conveyed through the event and participate in the meetings. On the occasion of the World Water Day on March 22, 2019, Parmarth workers carried out rallies in various areas of Bundelkhand to spread awareness among people for water and its conservation. The rallies saw participation of the youth as well as students in large numbers.

Other activities: In addition to the above-mentioned activities, Parmarth also organises other programmes and events to ensure a regular flow of communication with the different stakeholders in its water management initiatives. The common among them include workers of Parmarth reaching out to various villages and preparing work plans for identifying water-related issues with the villagers through social, economic, and natural mapping using their local knowledge, holding informal talks with them to know about their issues, organising singing programmes by mobilising the women of the villages, and others. Women from various villages associated with Parmarth have prepared songs in their local dialects giving messages of conserving water, forests, etc. which they sing occasionally to mobilise their communities from their village. Some of the lyrics of such songs are: *Kuda kachra na dalo pani mein, sukh chane agar zindgani mein* (don't dump garbage into water bodies if you want happiness in life), *Ore saajan kaato na ped, panchhi let basera* (Oh dear! Please don't cut the tree as birds take shelter on it), etc.

The Parmarth officials, through informal interactions with the local men and women in various villages, also motivate them to demonstrate and agitate before the government officials for the redressal of crisis-like situations that they face due to the lack of water resources. In recent times, Parmarth has led several such agitations by the villagers who accompany them to local government offices where they submit a memorandum to the concerned authorities like collectors or sub-divisional magistrates (SDMs) to get their issues resolved. This way, it facilitates communication between the communities and the government and administration. In 2017–18, when the president of the Machhuara Samiti (fishermen committee) of Bada Malahra block of Chhatarpur, Ajuddi Raikwar informed Parmarth representatives and volunteers of the district about the lakes going dry that caused a crisis-like situation among the community. They planned to stage a demonstration in front of the district headquarters in Chhatarpur. When the first demonstration did not yield any results, the community members organised subsequent rallies led by the district coordinator of Parmarth and its volunteers. Finally, after four demonstrations by the *Machhuara Samaj* (fishermen community), as Ajay Yadav informed, the district administration initiated action and ordered the deepening work of the lakes and ponds of the area so that they could store more water during the monsoon and hold ample amount of water even during the following summer season. This helped the community members to sustain their livelihood as the main source of earnings is fisheries in the community lakes and ponds. The initial work carried out by the district administration did not benefit the fishermen communities of the area as the job was given to the contractors by the district administration who carried out the deepening and widening of the ponds without any consultation with the villagers. Later, on the complaints of the Parmarth officials and the fishermen committee members, the deepening and widening was carried out by involving the local communities. The monitoring of the project by the local fishermen on the basis of their local knowledge and requirements helped the cause.

Besides this, the organisation also conducts programmes to mark special days like Environment Day, Women's Day, World Water Day, and others where it invites all the stakeholders including communities from different villages, activists, government officials, local organisations working in the sector, grassroots level workers, etc. and recognises their efforts. During such programmes, people doing outstanding work at different levels including grassroots level volunteers, community members, and activists are also felicitated for their contributions to bringing about a change in the status quo to encourage others to participate in the process of social transformation in whichever way they could. A similar programme was organised by Parmarth on March 8, 2017, at the Government Girls' PG College in Chhatarpur to mark International Women's Day which saw the participation of women from all sections of society in large numbers. Women officers from the government and administration, women activists, *Jal Sahelis*, women social workers, and several others from different walks of life were invited to the programme to share the stories of their success and empowerment, so that hundreds of women attending the function could draw inspiration to become the change-makers in their society. Many of the women were felicitated by Parmarth for their contribution to bringing about various positive changes. *Bund Bund* (drop-drop) being the theme of the function, many women talked about the water-related struggles they had to face on a daily basis while others talked about the initiatives that they had taken up around the issues of water either on their own or with support from Parmarth and the local administration. The *Jal Sahelis* from different villages of districts like Jalaun, Lalitpur, and Hamirpur in Uttar Pradesh and Chhatarpur in Madhya Pradesh too shared their experiences at the event and urged women to come forward and take action to address the water-related issues with collective efforts.

Water-shed Development in Chhatarpur District of Madhya Pradesh

Parmarth has implemented water-shed development projects in ten villages of the Bijawar block of the Chhatarpur district under the Rajiv Gandhi Watershed Management Mission (RGWMM)[1] of the Government of Madhya Pradesh. Parmarth is one of the organisations among the 53 partner

NGOs listed for implementation of the programme in different districts of Madhya Pradesh. As a part of this programme, Parmarth has constituted village-level water-shed committees. The committees that generally have 10–12 members are headed by one president who is elected through voting by the other members. And then, all activities related to water management including, the construction of check-dams, stop-dams, farm ponds, etc., are carried out in consultation with the members of the committees.

These committees play an important role in mobilising the villagers at the local level to garner their support and contribution to the work. The villagers as well as the members of the committees together contribute their part in the projects either in monetary forms or they offer *Shramdaan* (voluntary contribution of labour) during the construction of the structures for water conservation.

During the visits to Diha and Vaksuhi villages of Chhatarpur and after the interactions with the presidents and members of the water-shed committees in these villages, it was found that one stop-dam[2] has been constructed in Diha village while four stop-dams and one farm pond were constructed in Vaksuhi village of the Bijawar block. The members of the committees informed that these structures have made water available to them for irrigating their farmlands, which helped them sustain their livelihoods and prevented their migration to urban areas for work. The state coordinator of Madhya Pradesh, Rupak Ghosh, stated that the village-level water-shed committees (VWCs) are crucial in terms of post-project operations, maintenance, and future development of the assets created during the project. During a visit to the Panwari village of the Ghuwara block of the Chhatarpur district where Parmarth had constructed a check-dam[3] earlier and installed a sluice valve, it was found that a similar village-level water-shed committee called *Pani Panchayat Samiti* was functioning to maintain the quality of the structure.

Women as Mobilisers and Communicators

In parched Bundelkhand, women and girls are always at the receiving end of the water scarcity caused by frequent droughts. Thousands of young girls cannot go to school as their entire day passes in managing water for their homes. Looking at their hardships, Parmarth decided to mobilise women in the villages to come forward to work towards addressing the problems of the water crisis. As a result of this, it has so far mobilised hundreds of villages under its initiatives to form *Pani Panchayat* (water collective) at village level which is led by women. The women in the leadership roles in these Panchayats are recognised as *Jal Sahelis* (female friends of water). Over the years, the women have developed remarkable communication skills and have played a significant role in mobilising other women in the villages to join their efforts to secure water for their day-to-day chores. And with their increased level of confidence and sense of empowerment which they gathered after stepping out of their houses (as discussed at length in chapter 3), they have been equally successful in mobilising the support of the men in the villages and the government authorities. As a result of this, hundreds of new water conservation structures like ponds, lakes, check-dams, stop-dams, farm-ponds, etc. have been developed and the old ones have been renovated with collective efforts.

As a part of their communication strategies to garner support for their work, they make door-to-door visits and talk to the women for coming together and address the issue at hand. Many times, when their husbands or mothers-in-law and fathers-in-law would object to their idea of "women moving out of the houses," they would hold lengthy dialogues with them as well and try to convince them about the need for doing so. In most of the cases, they would succeed. Geeta, the president of the *Pani Panchayat* of the Manpur village the Babina block of Jhansi, says,

> Even if we fail to convince the husbands and other in-laws in one go, we make several attempts, and in most of the cases we succeed sooner or later. She said that there have been many instances wherein initially husbands and the mothers-in-law raised objections. But later when they realised that their coming together could ensure availability of water within the village, they not only allowed their wives and daughters-in-law but they themselves joined hands for the endeavour. That is how we desilted the pond in our village that could not hold water for long, after the monsoon.

She informed that they themselves had revived the pond by removing the silt, deepening it, and expanding its size, which ensured water for irrigation and livestock for a whole year. And for drinking water, she informed that with their demonstration at the district administration office, a huge overhead tank was established in the village which would be supplied with water through a motor fitted below it and taps were installed there so that people could get drinking water. Geeta said that even now, some women had to walk for around half a kilometre to collect water from there but it's a great respite as they do not have to walk two kilometres like they did earlier to get water from another village where, at times, they had to wait for hours for their turn to come. However, she said that the motor installed below the tank to pump water into it on the top has gone defunct, and it was not working for the last 15 days. She said that they have complained about it to the concerned authorities, but nothing has happened and they were planning an agitation at the district administration office soon.

The secretary of Parmarth, Sanjay Singh, said that right from the beginning of the organisation their focus was on uplifting women. He further said:

> We (I myself and four other friends who joined hands to work for the society) would see the struggle of women in terms of procuring water and the effects of the water crisis on women and girl children. Initially, we thought of educating the women and helping the school children go to school to get education. While we started working for sending girl children to the school by talking to their parents, we also made efforts to educate women in the villages thinking that it would help them in living better lives. However, the challenge was that we would hardly find women and girls who would be at least tenth pass so that we could motivate them to educate other adult women. There were villages where we could not find even a single girl or woman who studied up to standard tenth. Even though we were disappointed with this, we were not willing to give up, and strengthened our resolve to work for our mission.

Singh further said,

> As we were trying to work in this direction and making a slow headway, we came in contact with a collector in Jalaun, Rigzin Samphel who was honest and wanted to team up with some youth to work for watershed development in the region to address the water-related issues. He wanted us to mobilise people at the village-level so that some water conservation effort could be initiated in collaboration with the local community members.

As a part of that work, new structures like ponds, check-dams, and stop-dams were developed, he informed:

> We saw that the new structures that helped people have water in their wells and in the hand-pumps which were in dried state for a long time had brought huge respite to the women and girls who had to travel long distances for fetching water. This inspired us to shift our focus to engage with work that could address the issue of water scarcity. That project was a great learning for us which gave us new objectives and insights," he said adding that "We continued the work of educating women but this time the objective was to mobilise them as well as the men of the villages to come together and start working for addressing water-related issues. We found this work more meaningful as it drastically reduced the workload on the women and girl children could also go to school."

In the process, he said that they had facilitated hundreds of meetings and dialogues between the community members and government officials. Sometimes, the villagers were more understanding and supportive and would come along immediately, while on other occasions they had to go to them several times to make them join their efforts to work for the redressal of their everyday struggles for water. In some cases, they would find good authorities who would be happy to lend an ear to their concerns and extend administrative support while in others, they would have to hold demonstrations and agitations to compel the authorities to act.

Singh said,

> There were several challenges and each challenge had a different basis, mostly deeply-rooted in the society, like patriarchy, strongly maintained caste-hierarchy, and feudalism. However, Parmarth, with the support of its workers and volunteers kept learning and finding ways to deal with all the challenges and worked tirelessly for achieving its goal to make the villages water-sufficient.

On being asked how he rates the success of Parmarth in helping the villages deal with the water woes, he expressed, "I think, we have only been 50 per cent successful in achieving what we wanted to achieve. But yes, we are learning with every effort that we are making. And we will keep doing so for as long as we can." According to him, limited resources, trained officials switching jobs, administrative and bureaucratic challenges, feudalism and caste structures, plus his own frequent travel for national-level events to press for water-related policies and bills for water security for all affects the efficiency of the work in Bundelkhand.

WADI Project: Protecting Ecology and Enhancing Livelihood

In addition to the work related to water management, Parmarth is also involved in other activities like efforts to restore the ecological environment and promote sanitation and cleanliness among the village communities, to have kitchen gardens and fruit orchards in the backyard, and many others. As a part of its vision to promote other environmental-friendly activities at the grassroots level among the rural communities, Parmarth also undertook a project called WADI that was aimed at enabling the tribal people of the Lalitpur district to develop fruit orchards in their agricultural fields. They simultaneously cultivated vegetables in their farmland via intercropping method that resulted in enhanced income for the beneficiary families. As a part of this project, several new wells were dug and old ones were recharged for securing water for irrigating their fields. Besides, farm-bunding was also done to conserve rainwater for irrigation during the rainy season. Over 10,000 fruit trees were planted under the project which led to soil conservation in the intervention villages. This has resonance with Cox (2007) that the field of environmental communication must respond to the environmental condition in ways that benefit both human communities and the natural ecosystem.

This section of the chapter would discuss one of the projects being carried out by Parmarth with financial support from the National Bank for Agriculture and Rural Development (NABARD) which helps families from the Sahariya tribe with less than five acres of landholding in the Talbehat block of the Lalitpur district of Uttar Pradesh in developing orchards of fruit trees like guava, lemon, and amla (Indian gooseberry) along with forest species. A total of thousand such families were identified in Talbehat cluster of Lalitpur. The average financial support given to each family was little over Rs. 40,000 while the total cost of the project was around Rs. 4.41 crore. The project was implemented in 32 villages in the region. As a part of this project, many water conservation assets, like wells and ponds, were also developed in many villages, to ensure availability of water for irrigation in these orchards.

The project coordinator for this initiative, Shahid Ali, informed that the WADI project was started as a part of the NABARD's programme called Tribal Development Fund. The term "WADI" has been taken from Gujarati language which means small orchard. Ali informed that in addition to the fruit trees in the orchards, the families were also provided with the seeds of peas to be intercropped in the orchards which would give them additional income and enhance their economic condition. Besides the idea of environmental regeneration, the project coordinator stated that the project also aimed at initiatives like soil conservation, moisture conservation, development of water resources, enhancing communities' health, and women empowerment. He informed that so far a total of 954 orchards have been developed which have raised the economic condition of the families in the villages as they are able to sell the fruits like lemon, guava, and amla during their respective seasons along with peas. According to Ali more than 350 families practised intercropping pea in their orchards which helped them generate better income.

Manvendra Singh who is based at Parmarth's Talbehat office and was previously associated with the project as project coordinator informed that to make the project participatory, they had to hold several meetings in the villages with both men and women of the communities. Singh said,

> We would tell them about the benefits of developing the orchards and spending time in caring and nurturing it. Not all of them would agree to take up the work as they would work as daily wage labourers in the urban areas and would not want to invest themselves in any new project. But we convinced them that they would not have to work for the orchards every day and they could be working as daily wagers somewhere and still continue to work for the development of the WADI, sparing small amounts of time from their daily work schedules. As soon as some of them agreed, and took up the project, the others too followed.

He said that women were more eager to take up this work as they thought that if they could develop the fruit orchards their husbands would have the work of their own and the water resources developed for this purpose would provide them with water too. This way they would also be able to cultivate their farmlands and the husbands would not have to go out to work. So, many women played an instrumental role in convincing their husbands to take up the WADI project. He added,

> We also formed self-help groups at the village-level so that community members could come together and take up the responsibility for maintenance and sustenance of the orchards once the trees are planted. This way, the villagers would, sometime, themselves call to tell us about the problems they would face at the initial stage of the programme.

When the small plants started to grow and gained leaves and branches, animals would come and eat them. This was a big challenge that every farmer was facing. In their efforts to address this problem, Manvendra Singh said,

> When they apprised us about this issue, we talked to NABARD and they agreed to provide the wire for fencing to protect their orchards but they made it clear that the cost of getting them erected along all sides of the orchards would have to be borne by the farmers themselves. There were many farmers who installed the fencing as soon as they got the wires but there were still a few who kept the fencing wire lying in open as they would be busy with other work to earn livelihood thinking that the project is not going to give them any economic benefits in near future.

He said that in any initiative that they would take in the rural areas, there are people who would be very motivated to follow the advice and there are a few who keep showing their careless attitudes. But they were still convincing them to erect the fence as soon as possible with a view to bring more awareness among them and develop an eco-sustainable behaviour through information, teaching, and persuasion to solve a problem – a practice having strong resonance with what Abbati (2019) calls the multi-faceted "soul" of environmental communication.

Promoting Kitchen Garden Practices for Waste Water Management

Under the WADI project, the work for promoting kitchen garden practices within the premises of the houses was also carried out widely. Many of the community members who had sown seeds of the vegetables like brinjal, peas, bottle gourd, chili, tomato, papaya, etc. were reaping benefit. As a part of their initiative to promote kitchen gardens practices, the staff members of Parmarth would go to the community members and demonstrate to them how they could make use of the empty land outside their houses and the wastewater to grow vegetables without any additional efforts. They would ask them to identify the area outside their houses where the used water would flow. Since the soil in the area would remain always moist, there was no need to make any separate efforts to water the plants.

The project coordinator, Shahid Ali, during field visits to such villages, shared some interesting experiences that they had while meeting with people in the region to convince them to adopt the practice of kitchen garden. Ali said that the people from the tribal communities would happily agree to sow the seeds provided by them in their backyard and would also show them the areas in their houses where the used water from bathrooms and kitchens would flow. However, those belonging to the upper castes like Thakurs, Pandits, Lodhis, Yadavs, and others would not like the idea. "Most of them would say that how could we grow vegetables from the used water from the kitchen and bathroom and eat them. The water is dirty and to think of eating vegetables and fruits grown with the help of that water is completely unthinkable to them," he said, adding that "So, we focused on promoting the practice among the tribal communities only and as of now, around 200 kitchen gardens have been developed in Talbehat block among the tribal communities."

The work of promoting kitchen gardens was carried out in the Tikamgarh and Chhatarpur districts of Madhya Pradesh as well. Similar experience was shared by Ravi Tomar, district coordinator of Tikamgarh for the WASH project of Parmarth, who had told that people from the scheduled castes (Dalit) would easily get convinced to develop kitchen gardens and make efficient use of the used water from the kitchen and bathing areas while those from the upper castes would object to such practices considering the used water from home as dirty. "We would find it very difficult to convince the villagers belonging to the upper castes to develop kitchen gardens because of their notion of the 'used water being dirty' to grow vegetables and fruits. Many a times, we would have to give up when they would seem to be getting aggressive. So, our focus was on promoting the initiative among the people from lower castes. Some of those people who adopted the practice really found it beneficial. They not only grew vegetables to consume themselves but also sold it in the market, although such cases were less. And later, some of the upper caste villagers after seeing them grow vegetables successfully, themselves contacted us and sought seeds from us and developed kitchen gardens on their house premises," Tomar said.

During field visits to villages like Panya, Bhonyera, Ramtoria, and others in the Chhatarpur district, the authors met with one of the Parmarth volunteers, named Uttam Yadav who is a farmer and also owns an ashram where he has developed a Gaushala (cow-shed) with 50 cows. He said that he volunteers with Parmarth because he finds its work very meaningful and it gives him satisfaction to help him in their work. He added,

> *Chhatarpur jaise kshetra mein aake paani ki samasya ko dur karne ke liye kaam karna asaan nahi hai. Aur jitna maine inka kaam dekha hai mujhe lagta hai ki ye bahut punya ka kaam kar rahe hain. Yahi soch ke*

> *mai inse juda aur jitna ho sake mai inki madad karne ki koshish karta hun.*
>
> (It's not easy to address the issues of water in a region like Chhatarpur. The way I have seen their work, I think they are doing a very noble work. That is why I have joined them and try to help as much as possible.)

After visiting the houses in which seeds were given, it was found that the villagers had taken the seeds from Parmarth but they had not sown them. When asked why they did not sow them, the villagers said that they will sow it soon. The seeds were provided to them around a month ago. After further interaction, it was discovered that the seeds were supposed to be distributed to the villagers by Parmarth staff members in the villages but they did not even come to these villages. They gave the seeds to those villagers who they came across in Chhatarpur town. There was no follow-up on the work either. The villagers in possession of the seeds informed that they were still mulling as to where to sow the seeds as they were facing scarcity of water. Uttam Yadav informed that actually, the Parmarth staff were supposed to go to these villages and explain to them the process of developing the kitchen gardens in their houses where the used water from bathing and cooking areas would flow but that did not happen. They simply tried to manage the entire thing remotely which failed completely. However, the management of Parmarth was completely unaware of the issue as they would mention the work of kitchen garden development in these villages during their interaction. Yadav also shared the names of the staff members who were supposed to be doing this but did not take the pain of coming to the village. He said that if the top officials of Parmarth come to know about it there would be action against the responsible ones but the local staff have done water-related work in the region, and therefore, they share a good rapport with the villagers so this issue will never reach the head office as there lies an understanding between the villagers and the concerned staff members. He said that since both the villagers and staff members are equally responsible for non-action, such issues will remain undiscovered for the people at the head office. The volunteer said,

> In such cases, the staff members at the district level think that they have distributed the seeds to the villagers so they will go and sow them the way they explained and the villagers think that this won't make any big difference to their lives, so it is going to remain unnoticed and undiscussed while in documents and reports they will count these villages and villagers where kitchen garden work has been done.

This was a manifestation of how gaps emerge between the planning and implementation of the development initiatives at the grassroots level. The same villagers who were actively engaged in the development of water management structures like check-dams, stop-dams, ponds, etc. did not act on developing a kitchen garden at their own houses just because there was no proper communication between them and staff members in a sustained manner. "I have seen the enthusiasm with which the village people work when it comes to working for water conservation and developing new assets for the same, but they were completely apathetic towards the work of developing kitchen garden," he said. On being asked what the reason behind it could be, he made a very important observation,

> *Sir, is kshetra mein logon ko paani se zyada badi koi samasya nahi dikhti. Yahan par sabzi se zyada zaruri inke liye paani hai. Isiliye jab paani ki baat aati hai toh ye turant kuchh bhi karne ko taiyar ho jaate hain. Paani na hone se in sabke khet sukhe rah jaate hain. Agar paani mil gaya toh sabzi to ye apne kheto mein hi uga lenge kaafi matra mein.*
>
> (Sir, in this region, people do not see any other problem bigger than the issue of water. For them, water is more important than vegetables. That is why they get ready to do anything when it comes to water. Their farmlands remain dry due to a dearth of water. If they get water, then they will grow vegetables in large amounts on their farms itself.)

He said that the workers of Parmarth should also have shown them examples of the people in different villages of Tikamgarh who have benefited from the kitchen gardens.

However, in the Makhanpura village of Bada Malhera block of Chhatarpur, the case was a little different. Around 10–12 villagers who got the seeds had sown them and were having vegetables like brinjal and tomato along with coriander available in their own backyard. Although they had grown them in a small area, the 20 odd plants of each tomato and brinjal had been sufficient to provide them vegetables for at least 3–4 times a week. Reeta Bai, wife of Ajrath Lodhi, informed that one of the staff members of Parmarth, Dhaniram, had come to their village and given them the seeds and told them the process of sowing them. She said,

> On his advice, we had sown the seeds in the area where we would generally wash hands before and after eating or on other occasions. This would keep the area moist and we did not have to make any special efforts to water these plants as they emerged from the soil.

On being asked if she found the idea of kitchen garden beneficial, she replied,

> *Ab sir bina kuchh kiye haemin ghar pe khud ki sabzi khane ko mil rahi hai toh faida toh hai hi. Hamara thoda paisa bhi bach jaata hai kyonki saptaah mein 2-3 din sabzi nahi kharidna padta hai. Toh hame taazi*

> *sabzi bhi mil jaati hai, paisa bhi bachta hai aur yahan paani ki dikkat hai toh paani bhi alag se nahi dena padta.*
>
> (Of course, it is beneficial Sir as we are getting to eat our own vegetables at home. We also save some money as we do not have to buy vegetables for 2–3 days in a week. So, we get fresh vegetables, save money, and since we face water scarcity here, we do not have to make separate arrangement for water for it either.)

Similar stories were shared by some other homemakers of the village like Usha Bai, Krishna Bai, Noni Bai, and Sukhbati who had developed a kitchen garden of their own at their houses.

In another village, Kabaiyan Kheda of the same block, a total of nine kitchen gardens were developed. One of the women of the village, Laddo wife of Kaluwa Raikwar, who had developed the kitchen garden at her house said that her husband and children love mixed vegetables of tomato, brinjal, and potato and now, they had to purchase only potatoes from the market while brinjals and tomatoes are available at their own house. During the season of brinjal and tomato, they will ask me to cook that vegetable at least 3–4 days a week, she said and laughed. Dhaniram Raikwar, district coordinator in the Chhatarpur district of Madhya Pradesh, said that he himself went to all these houses to distribute the seeds and explained them the process. In many cases, like at Kaluwa Raikwar's house, he said, he himself suggested the area outside their home and made Laddo sow the seeds there:

> Sir, whenever we talked to the women of the houses about the kitchen garden and asked them to develop it at their houses they did it very sincerely. They are more sensitive towards the needs of the family. When they realised that some money could be saved with their small efforts and they could have home grown vegetables, they liked the idea and adopted it immediately. And the fact that there was no need for arranging water separately for these vegetable plants, it became an idea even more convincing to them.

However, requesting not to mention the names, Dhaniram informed that some of the staff members who directly gave the seeds to the male member of the families and asked them to sow them at the houses for developing kitchen gardens hardly followed the practice. They simply took the seeds and kept it with themselves without even discussing about them at home, he said.

If we look at Parmarth's work of promoting the idea of developing kitchen gardens within the premises of the houses in villages, it primarily aimed at attaining three goals – making efficient use of used water at houses, meeting the nutritional needs of village population to some extent, and helping them economically. Looking at it from the perspective of environmental communication, one could find the project helped the community members especially women pay attention to using the spare land in their backyard and the wastewater for growing vegetables. This realisation by the women to make optimum use of the natural resources around them and the green recovery of the land through fruit trees in the orchards under the WADI project by the community could be linked with what Stern (2000) called an "environmentally significant behaviour."[4] Such interventions are also effective in building the understanding required in the society about the significance of proenvironmental behaviours to address environmental problems Stern 2000). It does not only shape the pro-environmental behaviour at the individual level but when other villagers see the practice they try to emulate it at their houses too which eventually leads to the development of a community behaviour that is beneficial to the environment in which they live. No chemical fertilisers are used to grow these vegetable plants. This prevents the soil from any kind of pollution and makes chemical-free vegetables available to the beneficiaries unlike those purchased from the markets. Such kitchen gardens were developed by many people, especially those coming from the tribal and Dalit communities in many other villages including Harinagar, Marikheda, Udanna, Purapatti, Bakpura, Jhigri, Sukhyal, and Hardaul Patti.

Conclusion

According to Pezzulo and Cox (2018), the field of environmental communication serves as both crisis and care disciplines. The frame of crisis discipline deals with addressing the pressing issues of the contemporary world like the extinction of species, climate change, deforestation, disappearing wildlife habitats, pollution, cancer, declining natural resources, etc. As a care discipline, the field deals with exploring the relationship between human and the non-human world and highlighting the aspects of interdependence and biodiversity. While analysing the function of environmental communication in the context of the Bundelkhand region, one would find that it comes under the frame of both crisis and care. As a crisis discipline, it brings to the fore the issues of drought, water crisis, women's struggle for water, agrarian crisis, migration, climate change, unemployment, etc. There are reports and studies by NGOs, development organisations, government departments, and media that prominently reveal these challenges of the region to the society. And as a care discipline, it aims at building stronger relationships among the different stakeholders including the local communities, men, women, government and local administration, and civil society groups to work towards addressing these issues as was seen in the case of Parmarth's work. Through planning and strategies, it has carried out targeted interventions that seek to address these issues by sensitising people and motivating them towards activities such as rejuvenating water resources like rivers and lakes, developing water-sheds, creating other water conservation structures like farm-ponds, check-dams, stop-dams, etc.

Communication about the environment and ecology through actions as underlined by Pezzulo and Cox (2018) and Corbett (2006) could be seen in Bundelkhand in many activities. These could be demonstrations before the local administrations, rallies and *yatras*, protest marches, etc. All these activities, at the end, lead to people's engagement with the issues of water, farming, livelihood, and others. They trigger discussions and deliberations among various stakeholders to involve more and more people in such practices. As activities for communication, they intend to shape people's understanding about the natural environment in which they live and their relationship with the nature which is what environmental communication is considered vital for as pointed out by Milstein (2009), Cox (2012), and Mishra (2013).

In many parts of Bundelkhand where such interventions are made by Parmarth and other civil society organisations, the coming together of the people thanks to the increased level of awareness around the issues of water and the resultant rise in the discourses related to water can be seen as the emergence of what Yang and Calhoun (2007) and Cox (2012) call a "green" public sphere around the environment in general and water in particular in Bundelkhand. The emergence of such a sphere could also be likened to the idea of UNESCO (1997) that "educated minds and instincts are needed not only in laboratories and research institutes, but in every walk of life" (p. 15). It must also be helpful in bringing people together and building relations among them that could help them realise the needs of the environment that they live in (UNESCO, 1997). The interventions by Parmarth and other NGOs and groups are contributing a great deal in educating people, in this part of the country, to understand the environmental conditions and devising ways to address the pressing issues that affect their lives. The rejuvenation and creation of hundreds of water conservation structures in the region with collective efforts is nothing but the result of such education and awareness initiatives.

The village-level committees formed in the villages of Chhatarpur district of Madhya Pradesh to bring people together to create infrastructures like check-dams, stop-dams, and farm ponds to deal with the droughts is one such example wherein people were informed and trained about the ways to deal with issues of water crisis. In the process, their traditional knowledge too was taken into consideration to understand the landscape and tap maximum rainwater during the monsoon season by developing the water conservation structures. So, the communication strategies for this project were not only to raise awareness among the local communities about the ways to deal with the issue at hand but also to reach out to them in their villages and harness their local wisdom before initiating any water conservation project. For all such projects, Parmarth would carry out various kinds of communicative activities like meetings at the villages with the communities, water rallies, marches, workshops, night *choupals*, door-to-door visits for mobilising women, cultural programmes in local languages like street plays, singing, etc., mediating dialogues between communities and the local administration, river walks, etc. These activities would help Parmarth officials connect with the communities effectively and send the messages across.

Building a network of women in the rural areas who suffered the most due to water scarcity and facilitating them to become the change-makers too was a noticeable case of grassroots communication. The constant efforts of Parmarth's staff members to mobilise the women around the issues of water resulted in women themselves becoming the mobilisers and communicators who started engaging with all stakeholders from the local Panchayat representatives to district administration to the chief minister of the state. The women identified as *Jal Sahelis* by Parmarth for their leadership roles in water management initiatives at the local level became a talking point of the region and their work was recognised by the district authorities as well as the chief minister of the state who felicitated them for their work and asked the government officials to promote such practices in other parts of Uttar Pradesh too.

Having experienced the work of Parmarth like the water conservation projects, WADI project, and initiatives to promote kitchen garden practices from close quarters, it could be stated that in remote and rural areas of a backward region like Bundelkhand, "NGOs are among the most visible sources of environmental communication" as also suggested by Pezzulo and Cox (2018).

The secretary of Parmarth, Sanjay Singh said that they were happy to see that their efforts had brought about positive changes in several pockets of Bundelkhand. However, they felt that for the long-term and effective solutions to the water crisis that was becoming a common feature across the country, there was a need to reach out to the varied and wider sections of the society. With this idea of mobilising various sections of the society on national level, Parmarth started a nationwide campaign to bring the members of the civil society organisations together so that they could effectively advocate for policies required to deal with the issues of water. This campaign was named *Jal Jan Jodo Abhiyan* (which translates to "water people alliance campaign"). The next chapter discusses the activities and efforts being carried out by Parmarth in association with other civil society organisations and concerned individuals under this national-level campaign.

Notes

1 Rajiv Gandhi Watershed Management Mission (RGWMM) is a programme started by the Government of MP in 1994 for improving land and water resources in environmentally degraded villages of the state. The programme's official document mentions the larger goal is to move water-shed management from being a techno-centric programme to a people's movement through participatory bottom-up approach.

2 A structure built over a monsoon stream or drainage channel to block the rainwater run-off. The reserved water this way is used for irrigation.

3 A structure built over small rivers, streams, or monsoon drainage channels to reduce the velocity of the flowing rainwater. This helps in percolation of the water under the ground recharging the groundwater table in addition to allowing the water to flow downstream.

4 Stern (2000) in his article defines environmentally significant behaviour as one developed with an intention to affect some change in the environment usually in a beneficial way.

5 Campaign for Collaborative Action

Advocacy by Civil Society about Water Matters

"The need is not to link one river with another but to connect people with water bodies."

– Rajendra Singh, "Waterman" of India and Stockholm Water Prize Laureate[1]

Water is one of the most crucial natural resources that directly affect our day-to-day life. In the last few decades, the availability of this natural resource has been getting scarce for several reasons like – climate change, pollution, rapid urbanisation, industrialisation, rising population, etc. And this unwarranted change in the water scenario across the globe has led to tremendous struggles and conflicts not only within a local region or state but also across national and international boundaries. A perusal of some of the scientific studies and reports suggests that if no effective and immediate measures are taken, the concerns around the availability of water will only keep aggravating with the ever-rising demand for it both for commercial/industrial and domestic purposes (Bigas, 2012; Pacific Institute, 2009; Parikh, 2013).

Although two-thirds of the planet is covered with water, only three per cent of the total water on earth is freshwater, out of which two per cent is frozen in ice caps and glaciers, and thus, only one per cent is available for human consumption, in the forms of lakes, ponds, rivers, streams, etc. (IDSA, 2010). India has only four per cent of the total renewable water resources available on earth even as it is inhabited by 18 per cent of the world's population (POMIO, 2014). Besides this, the growing population, increasing urbanisation and commercialisation of agriculture, and a lack of judicious management of water resources have exasperated the problems of scarcity and degradation of water resources in India (Singh, 2000). Owing to the fast-paced development activities, there is relentless exploitation of the available water resources which is resulting in decreasing per capita availability, and a significant gap between the demand and supply of water in the country (Kumar et al., 2005; EBTC, 2011; UNICEF et al.,

DOI: 10.4324/9781003350286-5

2013; Bowonder, 1987). India being an agriculture-based nation, irrigation accounts for the highest consumption of water from the available resources in the country (Kumar et al., 2005; EBTC, 2011). As such, agriculture as a sector is also facing many challenges due to the dearth of water in many drought-prone regions of the country. Besides this, the increasing consumption of water by the industrial sector and the disposal of waste by the industries into the water bodies are further adding to the issue of water shortages in many parts of the country.

To address the issue of the frequently arising situation of water crisis, Shah and Koppen (2016) stress the need for honing the social capacity of communities and societies to adapt to water scarcity. Given the present scenario in India, Shah and Koppen (2016) call for the need for "a transition from water resource development mode" to "water resource management mode by embracing Integrated Water Resource Management (IWRM)" (p. 16). For them, while the former focuses mainly on the supply side, the latter aims at managing the demand aspect in a way that ensures sustainable use of the available water resources (Shah & Koppen, 2016). And when it comes to IWRM, civil society associations in the form of local voluntary organisations and NGOs (Non-Governmental Organisations) could be effective. Parameswaran Wishart (2000) feels that NGOs could play a crucial role in mobilising communities, building their capacities and extending support to the communities in activities related to water management.

In the past few decades, India has seen many noticeable efforts from civil society in the water conservation and management sector. A large number of NGOs and civil society organisations have taken initiatives to revive the traditional methods of water management for both irrigation and domestic use purposes, like rejuvenating *naula-dhara* (springs and streams) in the hilly regions of Uttarakhand, *bawdis* (step-wells) in plains, *johad* (rainwater storage tank) in Rajasthan, *ahar* (a rectangular embankment-like water harvesting structure), and *pyne* (artificial channel to bring river water to the fields) in south Bihar region (Koul et al., 2012). The organisations working in the water management sector in India are quite diverse in terms of the nature of their interventions, size, resources, areas of expertise, etc. There are big organisations like the Centre for Science and Environment (CSE), Paani Foundation, and the Energy Research Institute (TERI) which collaborate with NGOs across the country on the issues of sustainable development and natural resource management, through capacity building, research programmes, and also by serving as public interest and advocacy organisations. There are also many other organisations like Himalaya Seva Sangh, Rashtriya Jal Biradari, Paani Foundation, Pravah, Tarun Bharat Sangh, and others to name a few that work on water and sanitation programmes at the community level in rural and urban areas in their respective regions.

The Ministry for Water Resources, River Development and Ganga Rejuvenation of the Government of India (now, Jal Shakti Ministry) too felt that the issue of water scarcity needs all-round efforts and that the state could not cope with the problem on its own. Thus, one of the five goals, as mentioned in the National Water Mission document, is "promotion of citizen and state action for water conservation, augmentation and preservation." And one of the recommended strategies to achieve this goal is to "encourage the participation of NGOs in various activities related to water resources management, particularly in planning, capacity building and mass awareness" (Ministry of Water Resources, 2011, p. III & IV). These goals continue to be part of the National Water Mission of the newly formed *Jal Shakti* Ministry of the Government of India. In 2019, such groups and individuals were felicitated by the ministry for their remarkable work in this direction.

So, the potential of civil society organisations as important actors in the field of water management is acknowledged not only in academic scholarship, but the governments also look at them as essential partners in the implementation of their programmes and vision. It is a widely acknowledged fact that civil society plays a vital role in bridging the gap between the governments and the society as they work directly with communities and are associated with vast pluralistic networks in different regions of the country. In India, civil society organisations have been instrumental in helping the government implement its programmes and schemes at the grassroots level even in remote areas. Thousands of NGOs are engaged in taking the work of the government to remote areas and are implementing them with the support from the local government authorities as well as the communities and local representatives. Civil society organisations are also engaged in activism and advocacy, demanding action and policies for the common good of society. This chapter analyses in depth, the Jal Jan Jodo Abhiyan (JJJA) (water people alliance campaign) as a civil society–led campaign for water as it incorporates in its communication strategies, the components of both activism and advocacy for conserving water and seeks to ensure water rights of the people. Before examining the JJJA, its approach, and processes, the chapter in its second section attempts to comprehend theoretically the multiple operationalisations of the concept of civil society and the nuances of its manifestations and functions through a reading of the available literature. The third section of the chapter explains Parmarth's strategies to bring people working in the water sector in different parts of the country together and mobilise them to make collaborative efforts for ensuring water security for all citizens in the country by protecting the water resources. This section particularly captures the initial thought processes and churnings that happened at Parmarth as they envisioned the JJJA and the following consultations with the relevant actors in different regions. The fourth section discusses the campaign's endeavours to create literacy and awareness among the people around the issues of water to seek their participation and motivate them to take action on their part to protect the water bodies. The underlying strategies prepared by the campaign leaders and members during different events to spread water literacy and awareness are also shed light on. The next part of the chapter analyses the advocacy efforts made under the campaign for convincing the governments for bringing a water security law and the subsequent processes of dialogues and negotiations with representatives of the governments both at the centre and the states and the members of parliament and legislative assemblies. The sixth section is the description of how the consistent efforts of the JJJA campaign leaders led to some success as the Congress-led Madhya Pradesh government that came to power in the year 2018 decided to draft and introduce the water security bill in the state assembly. The last section concludes the chapter by synthesising the previous sections and pulling together the significant points from them.

Civil Society: Definition, Structure, and Functions

Several scholars regard civil society organisations functioning at different levels – local, national, or international – as significant agents of change affecting people's lives in various ways (Khilnani, 2002; Cohen & Arato, 1994; Mahajan, 1999). The domain of the civil society's work globally ranges from ensuring equity and justice for the marginalised to promoting citizenship and democracy and pluralism (Vinod, 2006; Beteille, 2001; Mahajan, 1999). There is evidence to show that civil society organisations have made distinct contributions in the area of social change (Edwards, 2004). In India too, the civil society constituted by voluntary associations, community-based groups, and non-governmental organisations (NGOs) has been playing a valuable role in addressing contentious issues and carrying out grassroots work in distant places with communities that are marginalised or deprived (Chandhoke, 2007; Giri, 2012; Das, 2007).

Alice Bouman-Dentener of "Women for Water Partnership" and Bart Devos of "World Parliament on Water" during the UN-Water Annual International Zaragoza Conference – 2015, described civil society as an arena where people come together to advance the common interest because they care enough about something. They added that civil society had played a key role in complementing the work of state actors and intergovernmental organisations. Similarly, Anil Bhatt (1995) notes that the voluntary organisations, in their limited area of operation, have been quite effective as they have made a significant impact, particularly in those areas which are considered soft and complex, and where large legalistic and bureaucratic organisations have not been able to make any breakthrough. Sunil Khilnani considers "civil society" as an entity that throws up "inputs" and "demands" which the state must service and accommodate failing which would mean a "crisis of governability" (Khilnani, 2002).

Gurpreet Mahajan (1999) emphasises on the constructive relationships between state and civil society and says that in a democracy, the state and civil society work together, in collaboration with each other, for the promotion of citizenship rights and the conditions of subjective freedom. She calls this link between civil society and the state as the defining feature of a democratic system of governance. In the modern context, this idea of collaboration between the state and civil society organisations as suggested by Mahajan (1999) appears to be functional as we find several voluntary organisations being actively involved in various sectors ranging from health, livelihood, environment, sanitation, education, and the others. As of now, several projects are being carried out wherein the governments facilitate the NGOs to execute development work in specific sectors. However, in many cases, civil society organisations take up various projects on their own for people's welfare by having funding sources other than the government.

Benjamin R. Barber (1995) considers civil society as the domain which facilitates the mediation between the state and private sectors and offers women and men, a space, where they can carry out both voluntary and public activities. For Barber, civil society is a space that unites the virtue of the private sector – liberty – with the virtue of the public sector – concern for the general good. He calls the formation of civil society "voluntary, and is constituted by freely associating individuals and groups. But unlike the private sector, it aims at common ground and consensual, integrative, and collaborative action. Civil society is thus, public without being coercive, voluntary without being private" (Barber, 1995, p. 114).

Neera Chandhoke (2003) thinks of civil society as a "third sphere" other than the state and the market irrespective of its constituents like – associations, voluntary agencies, or social movements, while some also consider civil society as an alternative to the state equating it with non-state and non-government organisations (Vinod and Deshpande 2013). Andre Beteille (2001) seems to be re-establishing the idea of civil society and state being complementary to each other when he says that civil society shrank and withered in the regions where the state was hostile to it like in the USSR under Stalin or in Germany under Hitler. In contrast, in countries like Britain and the Netherlands, where the state was supportive, civil society grew and prospered (Beteille, 2001). Beteille feels that the state also needs support from civil society for its own health and well-being.

Collective action by the people in search of good societies has been a universal process of human experience as reflected in numerous ways across time, space, and culture (Edwards, 2004). Edwards articulates that the idea of civil society offers a significant emancipatory potential, explanatory power, and practical support to problem-solving in both established and emerging democracies. Today, for Edwards, civil society seems to be the "big idea" on everyone's lips – government officials, journalists, funding agencies, writers and academicians, and millions of people across the globe who find it an inspiration in their struggles for a better world.

For Anant Giri (2012), civil society refers to that conscious and mobilised aspect of the society which attempts to create a space for critical self-reflection and public deliberations. However, he also observes, "Civil society is also nourished by the support from the market. Contributions from actors from the market, such as corporate leaders and other market leaders, contribute to the resource base of civil society" (p. 229). Cohen and Arato (1994) in their book *Civil Society and Political Theory* while examining the roots of civil society in works of scholars like Arendt, Habermas, Foucault, etc. and applying them to the modern democracies understand civil society as a "sphere of social interaction between economy and state" which is composed above all of the intimate sphere (especially the family), the sphere of associations (predominantly voluntary associations), social movements, and forms of public communication. Cohen and Arato (1994) also assert that modern civil society comes into being through the forms of self-constitution and self-mobilisation.

Hence civil society is a contested concept both in terms of its composition and its activities. Most scholars, however, consider it as a non-state, non-corporate entity or a third independent space for deliberations and collective action. It is also debatable that if civil society were looked upon as complementary to and dependent on the state, then does it still remain an active site for social movements and independent thinking.

According to a *Down To Earth* report (2015), there are around 40 lakh registered non-profit organisations in India, out of which around one per cent are working on issues related to environment. This means approximately 40,000 non-governmental organisations are working in the environmental sector in India currently. In the last few decades, environmental concerns have been one of the major areas where civil society organisations have been making significant contributions in India (Gemmil & Bamidele-Izu, 2002; Princen & Finger, 1994). Organisations working in the environmental sector or environmental organisations have been instrumental in mobilising people around several environmental concerns through environmental movements and advocacy practices using different communication media (Dreiling et al., 2008). The ever-increasing water crisis across the globe is one of the most threatening among environmental concerns, and there have been many movements and conflicts around water between communities and the state. India too has seen social movements around water in the forms of Narmada Bachao Andolan and Tehri Dam movement, among the prominent ones.

Many civil society organisations in India are working in the water management sector to deal with the issue and ensure water security for communities facing water shortages in various parts. In addition to taking up several projects to overcome environmental crises, civil society organisations have also been instrumental in communicating and educating people about the existing concerns and in many cases also engaging with different actors to find solutions.

Drawing from these scholars, this chapter discusses the various events and activities that are carried out regularly under the JJJA and how these are helping them bring multiple stakeholders together to advance the debates for better laws from the government for addressing water-related issues and actions from the society as a whole. The nationwide campaign was started by Parmarth to advocate for a law that guarantees water security for all on the lines of the Food Security Act and to mobilise people across the country to contribute to the rejuvenation of the water bodies in their areas which are drying up for various reasons.

Towards Collaborative Action

For an NGO working in a specific region to think of launching a nationwide campaign to mobilise people around water needed innovative strategies. So, when they thought of the idea, they felt that it was necessary first to hold consultation with the people actively working in the water sector. Having worked in the area of water conservation for more than a decade, Parmarth was also connected with other civil society organisations and activists working in different parts of the country. One such organisation was Tarun Bharat Sangh founded by Rajendra Singh known for his exceptional work done in the water conservation through traditional water harvesting methods in the Alwar district of Rajasthan. Singh is also popularly known as the "Waterman of India." The secretary of Parmarth, Sanjay Singh knew him personally as they had met on different platforms as both of them were working in the water management sector. So, Singh decided to hold a consultation about the proposed campaign with the "Waterman of India" and seek his support. He said that the most significant advantage of having Rajendra Singh on board for the nationwide campaign was that he was very well connected with the activists and other people and organisations working in this area throughout the country. He would routinely be invited by the people for guidance and consultations in various parts of the country on the issues related to water.

Sanjay Singh said that he had also spearheaded several marches and protests in different parts of the country. Therefore, he thought that the "Waterman" could be instrumental in bringing like-minded people together for the nationwide campaign. On being approached, Rajendra Singh immediately gave a green signal saying that there was an immediate need for such a campaign in the country so that people doing working for water conservation in different parts could join hands for a larger cause. Following this, Singh said that the "waterman" was made the face of the JJJA as he agreed to lead the campaign. Following this, Parmarth and Tarun Bharat Sangh (TBS) contacted water experts and activists from various states and invited them on board for launching the campaign.

The campaign was launched in 2013 at Lucknow in Uttar Pradesh. And Sanjay Singh was made the national convener of the campaign on the advice of Rajendra Singh, and its national secretariat was established at the Jhansi office of Parmarth in Bundelkhand. The primary objectives of the campaign included (a) pushing for a water security law throughout the country which would ensure access to clean drinking water for the people of all sections of the society and (b) protecting rivers, lakes, and ponds through community participation by mobilising the people in various parts of the country through water literacy activities. It also aimed at offering a platform to the concerned individuals from all walks of society including grassroots water activists, students, scientists, academicians, students, lawyers, journalists, writers, government officials, and others to share their views and ideas about protecting the water bodies of the country to deal with the issues of droughts and water crisis. Through this campaign, Parmarth also wanted to strengthen the voices of the deprived sections of the society such as the tribals, Dalits, and women for securing water rights in the remote areas where they struggled a great deal for procuring water.

The "Waterman of India," who has also won Magsaysay Award and Stockholm Water Prize for his extraordinary work in water conservation, said that the main objective of the campaign, as its name suggests, was to connect people with the issues of water.

Sing said,

> Unfortunately, our governments are working towards connecting rivers through river-linking projects, which is very dangerous for both the health of the rivers as well as the ecology; whereas the need is to connect people with water who can understand the causes of the water crisis and work towards a solution.
>
> That's why the sole agenda of this campaign was to reach out to the maximum number of people and the governments and have them realise the threats that lie before all of us.

The campaign leaders are active in 22 states in India where they are working towards mobilising and strengthening civil societies on the issues of water.

Hundreds of big and small civil society organisations including NGOs, and CBOs joined the campaign from different parts of the country. Among those well known for their efforts include Dasholi Gram Swarajya Sangh (DGSS) founded by the Chipko Movement leader Chandi Prasad Bhatt; Ekta Parishad founded by Rajagopal PV; Rashtriya Jal Biradari a network of water activists started by Tarun Bharat Sangh; Sagar Mitra Abhiyan; Gandhi Peace Foundation; Manavlok – a Maharashtra-based NGO; Ghoghardiha Prakhand Swarajya Vikas Sangh (GPSVS); and many others. Similarly, many prominent people extended their support to the campaign that included politicians, administrators, social workers, religious leaders, activists, farmer leaders, academicians, and others. Some of those included Late Gyan Swaroop Sanand (Prof. GD Agrawal), Anna Hazare, Chandi Prasad Bhatt, Swami Agnivesh, Rajagopal PV, Gandhian thinkers SN Subbarao and Ramesh Sharma, former IPS turned politician PK Siddharth, and many others. All in all, the campaign was supported by around 1,200 civil society organisations from across the country that included NGOs, CBOs, farmers groups, colleges, universities, and research and management institutions.[2] Most of the training programmes for the thousands of activists associated with the campaign were conducted at Tarun Jal Vidyapeeth (Tarun Water School) at TBS Ashram in Alwar, Rajasthan.

This way, JJJA was able to partner with various civil society organisations from more than 22 states of the country, as well as individual activists who wished to join the efforts of the JJJA. Having developed a network of people from all walks of life, JJJA carried out activities and events like water literacy programmes, consultation camps for the activists and campaign leaders, river marches for more than 100 rivers, and regional and national water conventions to bring various stakeholders together. These events would facilitate dialogues about how water bodies could be protected, and a case for a water security law could be advanced. During these events, all the NGOs and CBOs would present the work that they have done in their respective areas in the field of water conservation to inspire others about the possibilities of work that could be done by other individuals and groups. Many groups and individuals would also be felicitated towards the end of the events as a part of the recognition of their work to motivate them to keep up the good work. Some individual activists were given the title of Jal Yoddha (water warrior) for their exemplary efforts in protecting and reviving the water bodies in their areas.

These events organised in various parts of the country also helped JJJA get recognition for its work in the water sector. As a result of this, many state governments like those in Madhya Pradesh, Uttar Pradesh, Rajasthan, Maharashtra, Bihar, Karnataka, and others sought consultations and advice from the campaign leadership for various water-related projects that they wanted to carry out in their states. They were made members of several committees and task forces constituted by these state governments for carrying out work around water. For example, the Government of Bihar partnered with JJJA to study the condition of the River Ganga from Patna to Farakka in West Bengal and sought the recommendations from them about the steps that it could take to improve the health of the national river in the state. Similarly, the Government of Madhya Pradesh included the JJJA leadership in the task force that it created for drafting the water security bill for the state in 2019, a step taken on demand of the JJJA itself.

The activities and events organised under JJJA aim at creating water literacy among people and bringing awareness among them about the issues related to water scarcity and the probable solutions depending upon the nature of the problem. According to the JJJA leader Rajendra Singh, the government schemes and projects around water would not be effective in attaining the desired results unless water literacy and awareness are created among the stakeholders. The literacy and awareness programmes would help the local communities realise their roles and responsibilities in the water conservation initiatives. Once they start participating in such programmes wholeheartedly and taking ownership, the desired change in the water scenario would become possible. With these ideas in minds, the leaders of the campaign give massive emphasis on the aspect of building water literacy and awareness-raising among people. For this, they design and carry out several activities and events in various areas to effectively send messages for the need for water conservation efforts at all levels among all sections of the society.

Creating Water Literacy and Awareness

During a national water convention organised at Bijapur in Karnataka in August 2017, there was a considerable focus on developing water literacy centres in every state. The national convention was organised after the culmination of 101 river yatras (marches) that were carried out by 15,000 members of the campaign for four months. The yatras covered 101 rivers from Kashmir to Kanyakumari and Guwahati to Goa. The purpose behind these yatras was to study the condition of the major rivers across the country and create literacy among the people residing near these rivers. The convention was inaugurated with *Kalash Sthapana* (urn installation), which is considered auspicious for starting an important event in Indian tradition. This urn contained the water from all the 101 rivers.

The final day of the event saw the adoption of a declaration called "Bijapur Declaration." One of the important agendas of this document was to create water literacy across the nation by setting up water literacy centres in every state and district. It was decided that efforts have to be made by the members of the JJJA from different states to establish such centres in their respective areas. The Waterman announced that the work of setting up such centres has already been done in Bundelkhand, Rajasthan, and Maharashtra where water activists were identified as *Jal Sevak* (water worker), *Jal Doot* (water messenger), *Jal Premi* (water lover), *Jal Yoddha* (water warrior), and *Jal Nayak* (water hero) depending on the nature of their work and were being given capacity-building training at these centres.

Rajendra Singh, during one of the interactions with one of the authors on the sidelines of the event, informed that giving such titles to the people working on water at the grassroots level and appreciating their work in public forums gave them a big reason to take pride in what they were doing. When the ministers and other important personalities felicitate them at such events, and their pictures appear in the newspapers, it also motivates others in their villages to join their efforts. This process of giving due credit to such activists working at the grassroots level and recognising their efforts has been one of the key factors behind whatever success the campaign has achieved so far.

Singh said,

> These recognitions motivate them to continue the good work in the field of water management and also involve others in their areas to scale

> up good practices. That is also one of the reasons they travel to distant places to be a part of such events and dialogues and share with others the work that they do in their regions

He further noted that spreading literacy and awareness about water was essential to help the cause of water. Besides other objectives, one crucial emphasis of the JJJA was to build the local capacities of the communities inhabiting the areas around water bodies like a river so that they could take ownership and protect the existing sources of water.

Singh reiterated,

> That's how we have rejuvenated seven rivers in western Rajasthan and created hundreds of water harvesting structures in the region. We have seen that happening in Bundelkhand with a large number of women taking up the responsibilities of protecting the water bodies and reviving them with Parmarth's interventions. So, if one could reach out to the local communities and tell them in the right manner about the problems and the solutions, they have immense potential to change things around. That's why water literacy is very important in the county

On being asked whether the government should also carry out such water literacy programmes, he said that the governments do everything on papers. Still, the harsh reality is that they have always been lagging in terms of the "actual willpower" to implement. In his words,

> We have met with several ministers of both central and state governments. They never say that they do not want to do anything that we demand. But they do not have that kind of network of people who could actually go to the people living in the countryside and interact with them on a regular basis. That's why we have decided not to rely on the government for anything and expand our own network for bringing about the change that we want with our own efforts. Many ministers sincerely wish to contribute to this cause. But, the bureaucratic system which has to implement the ideas are rigid and complex, which proves detrimental for their intent. Most of the government employees are comfortable working in their offices and do not want to move out. That's why in the remote areas many of the government schemes are actually implemented by the NGOs.

Probably this is why Bhatt (1995) feels that NGOs have made significant strides in those areas of the development process which are considered "soft" and "complex" where the large bureaucratic organisations have not been able to make a dent. NGOs as civil society actors play a significant role in ensuring people's participation through creative and innovative ways of messaging and providing required training and education (Bhatt, 1995). It was evident in Bundelkhand how local women would prepare songs in the Bundeli dialect to appeal to people to protect the water bodies from pollution and use water judiciously. Besides this, during all kinds of local cultural events in the areas of its interventions, Parmarth would have the local singers and performers sing songs with an appeal to create farm ponds, plantations, save water, etc. also make them announce about any water-related programme that they would organise in those areas.

As a part of this campaign, it was seen in Bundelkhand that Parmarth has partnered with several Panchayat offices and the local elected representatives, schools, and their teachers and students to make them stakeholders in spreading the literacy and awareness around water. In schools, they formed "Child Cabinets" wherein students would be made ministers with different portfolios like water, sanitation, health, etc. These children would take an active part in the activities related to spreading awareness around water conservation like rallies, plays, and marches to disseminate messages pertaining to saving water and protecting water bodies. Through interactions with the principals of the schools and the authorities from the local administration and district education officers, Parmarth also installed rainwater harvesting structures in various school compounds and Panchayat offices' premises. Following this, many individual residents of the nearby villages also developed rainwater harvesting systems in their houses which helped them recharge the water level of the water resources like hand pumps and wells.

The national convener of the JJJA, Sanjay Singh, informed that they would organise meetings with the local communities during river yatras that they carried out along 101 rivers of the country and the way people would welcome them at various places would be overwhelming. He further shared that:

> At most of the places, people would come to us, walk with us and would express their willingness to work towards the conservation of water resources. Everyone can easily relate to the issue of water. During several such interactions, the local people themselves would tell us the problems associated with the rivers and other water bodies, including the groundwater. And they would also know the steps that could be taken to address those problems. These yatras, thus, would serve as a stimulus for the people to engage with the issues of water and motivate them to contribute in whichever way possible for the cause. So, these river yatras helped us not only in creating literacy and awareness but also in instilling the desire among the people to work for protecting the water bodies. They also led to expanding our network.

During such two river yatras in Bundelkhand one – Ghurari river yatra – and two – Bachheri river yatra – the researchers found that before starting these yatras, Parmarth officials develop a complete roadmap of the yatra and what messages are to be given to the people during these marches. They would hold meetings in the villages along these rivers where a large number of people from the village communities would gather and interact with the participants sharing their knowledge about the condition of the rivers and work needed to revive/maintain their flows in future. And looking at the everyday struggle of the people in these villages, Parmarth decided that in addition to focusing on the community-led water management initiatives, it was also important to advocate before the government for a law that ensured an equitable distribution of water among all sections of the society. Sanjay Singh said that if the governments agreed for promulgating a water security law for all, the condition of the water bodies like rivers and lakes would also improve as governments would have to maintain them for providing water.

Advocacy for a Water Security Law

Having realised the need for a water security law after looking at the depleting water resources across the country and related people's struggle to procure water, the campaign leadership started the advocacy for it on a national level. In the process of pushing for the law, the campaign organised events in different parts of the country where people from all sections of the society including government representatives and leaders from the political parties gathered and held dialogues. And the campaign came up with several declarations like Patna Declaration, Jalgaon Declaration, Bijapur Declaration, and Khajuraho Declaration following the national-level water conventions that the campaign had organised at these places. These declarations would have summaries of all the brainstorming sessions that would be conducted during these events and would include all the recommendations for the required actions around water. They would be shared with the media for the public's engagement as well as with the political leaders and the government representatives so that these issues could be discussed in the assemblies and parliaments.

Having developed strong visibility with their frequent events, activism, and dialogues with different important stakeholders, JJJA representatives and leaders also get invited by the government bodies for various programmes that they organise on issues related to water. When they were invited during one such programme in 2014 called *Jal Manthan* by the Department of Water Resources, River Development and Ganga Rejuvenation, the representatives of JJJA emphasised on the need for a water security law and the promotion of practices of water use efficiency across the country. They also stressed on the need for a nationwide water literacy drive by ensuring the engagement of the youth and others working in the water conservation and management sector in India. Since the beginning of the campaign in 2013, the campaign leadership has held meetings with over 500 MPs and MLAs across the country and sought their support on the water security agenda for the country.

Later, the campaign also prepared a draft water security bill in consultation with the water experts, activists, scientists, water practitioners, academicians, and others from different states including Karnataka, Maharashtra, Bihar, Himachal Pradesh, Delhi, Rajasthan, Uttar Pradesh, Madhya Pradesh, and others. Inputs for the draft bill were also gathered from grassroots-level water practitioners in these states during the yatras and meetings. Regional meetings were organised by the campaign which saw the participation of the concerned people who shared their views and ideas for drafting the water security bill. More than 250 members from civil society organisations, social activists, academicians, researchers, and water practitioners participated in these regional meets and discussions. They gave their suggestions and recommendations for the bill.[3]

To give impetus to the campaign for water security law advocacy, JJJA also partnered with the youth organisations like National Service Scheme (NSS), National Cadet Corps (NCC), and Nehru Yuva Kendra (NYK) in Uttar Pradesh, Haryana, and Rajasthan. More than 10,000 youth members joined the campaign to advance its objectives. They were trained at different programmes organised at state levels wherein they were taught how they could communicate to the people about the issues of water. Following these state-level training camps for the youth members of the campaign, a special two-day training programme was organised for them at Mandi in Himachal Pradesh which saw the participation of the youth from nine different states. The leadership of the campaign also kept updating the draft of water security bill after the meetings with the community members in the rural areas, political leaders, religious bodies and their representatives, and students from various universities and colleges. Technical support and inputs were also provided to the campaign in this regard from various universities and institutions like IIT Mumbai, IIT (BHU), IIT Delhi, Dharwad University in Karnataka, Lucknow University, and others.[4]

In addition to these, the campaign also prepared content for the manifestos during various kinds of elections in the country like Lok Sabha elections, state assembly elections, and local bodies' elections and shared them with the various political parties. Seeing that the political parties were simply accepting their content for manifestos, but were not doing anything about them, they launched an attack on the major political parties like Bharatiya Janata Party (BJP) and Indian National Congress (INC) through media saying that they were not concerned about the serious issues like water and only superficially talked about them in their respective manifestos. This way, the campaign succeeded in having the issue of water security included in the INC's manifesto for the 2019 general elections.

For effectively attaining the objective of the campaign to advocate for the water security law, a core committee meeting would be convened every quarter which would include the "Waterman of India", its national and regional conveners. These meetings would review the progress of the campaign activities and prepare the roadmap for the further plan of action for the next quarter.

In the Bundelkhand region, in addition to doing advocacy for the water security law with both the state governments of Uttar Pradesh and Madhya Pradesh, the campaign with the network of *Jal Sahelis* and *Pani Panchayat* groups also helped the governments in realising the need for drought relief programmes in the region in various areas. With the interventions of the *Pani Panchayat* members and the *Jal Sahelis*, around 3000 families benefited and received drought relief packages.[5] The members of the *Pani Panchayat* groups and *Jal Sahelis* helped the families affected by the drought fill out their applications and place them before the concerned authorities at the Panchayat, block, and district levels. Their interventions also saw the installation of 42 new hand pumps which ensured safe drinking water for over 3000 families in 5 villages while many defunct hand pumps were re-bored and repaired.[6]

With the advocacy efforts of the JJJA for protection of rivers and water security, the then Ganga Rejuvenation Minister of the Government of India, Uma Bharti, had ensured in 2017 that a special team of experts would visit Patna in Bihar and the Farakka barrage area of the Murshidabad district in West Bengal. This team was set up to understand the issues facing the national river due to the barrage and the struggles of the people residing near it due to the issues like siltation and the soil erosion. Following this, Uma Bharti constituted a special committee to study the issues related to the River Ganga and recommend the steps to be taken to improve the health of the river in Bihar and West Bengal. However, later as she was divested of the portfolio in 2017 during a cabinet reshuffle by the Central Government, the project was kept on a back-burner.

Similarly, the leaders of the campaign also held several meetings with the chief ministers of Uttar Pradesh previously with Akhilesh Yadav and later with Yogi Adityanath. In 2017, after discussions with the chief minister (CM), Yogi Adityanath, the campaign leadership was assured that both the issues of Ganga cleaning and droughts in Bundelkhand would be prioritised by his government. Following these meetings, the CM constituted a high-power committee to study the condition of the River Ganga in Uttar Pradesh and the state of droughts in the Bundelkhand region so that the required steps could be taken by the government to address these issues. However, when asked about the current status of the committee and their recommendation in this direction, Rajendra Singh informed that the Government of Uttar Pradesh has done nothing in these areas and that those were all false promises.

Similarly, the JJJA campaign held meetings with the state government of Madhya Pradesh to inform them about the dying water bodies of the Bundelkhand region of the state and the consequent situation of droughts occurring every year in the region. The earlier government, led by chief minister Shivraj Singh, took a lot of interest in the work of the campaign. It sought the advice of the campaign leaders for the *Narmada Seva Yatra* (Narmada service march) that was carried out in Madhya Pradesh for 100 days with a view to creating a mass movement for protecting the Narmada River considered as the lifeline of the state. During one of his visits to Madhya Pradesh to participate in the Yatra in December 2016, as a part of the campaign, Rajendra Singh had said that it was very important to carry out plantation on both sides of the Narmada River to ensure its natural flow in the future. Later, chief minister Chouhan had announced that a huge tree plantation drive would be carried out on both sides of the river and the farmers owning the agricultural lands would be given compensation until the fruit trees matured.

The chief minister himself had admitted during one of the JJJA events at Bheekhampura in Alwar, Rajasthan, that the government alone could not solve the water-related issues and protect the rivers. He had come to the capacity-building training camp for the water activists at TBS Ashram. He said that he wanted to participate in it as a learner to understand how his government could work better in protecting the rivers. He expressed a strong desire to work with the campaign and invited the campaign leadership to be a part of the expert committee that his government was constituting for the revival and protection of the Narmada River and for addressing other water-related issues in the state, especially in the Bundelkhand region. Following this, JJJA joined the government's initiative, and hundreds of members of the campaign had joined the MP Government's plantation drive along the river in July 2017. Over six crore saplings were planted within a day with thousands of people taking part in the event. Later, in December 2018, the government changed in the state following the assembly elections, and the work on the project to protect the Narmada River from further deterioration through the massive plantation on both sides came to a halt.

Madhya Pradesh Water Security Bill Draft: A Breakthrough

Even as the government had changed in the state of Madhya Pradesh, the JJJA leadership kept meeting the ministers and other prominent government representatives. Many government authorities knew them because they were working with the campaign leaders even in the previous government. According to Sanjay Singh, one significant thing that JJJA leadership had done before the state assembly elections in 2018 was that they had the water security agenda included in the manifesto of the Congress party. And immediately after the formation of the new government in Madhya Pradesh, they organised a national water convention in Bhopal in which three ministers from the government participated, and they agreed that there was a need for a water security law in the state.

Singh felt,

> We felt that if the water security law is promulgated in the state, the government will become answerable to ensure access to clean drinking

> water to the people and to protect the water resources in the state many of which are dying because of the apathy of the government. Once the government decides to bring the law and becomes answerable, a lot of things could be done in the water sector. Even as the government would not be able to do everything on its own, at least they will be able to partner with hundreds of civil society organisations in the state and implement the water security agenda.

After frequent meetings with the JJJA leaders, chief minister Kamal Nath agreed to bring a water security law in the state to protect the dying rivers and their tributaries in Madhya Pradesh and address the issue of water shortages. The chief minister invited the leaders of the campaign to be part of the task force constituted by the government to draft the water security bill to be tabled in the assembly. The group of experts created for the drafting of the bill included the noted economist and the President of Bharat Rural Livelihood Foundation, New Delhi, Mihir Shah as the chair and Rajendra Singh as co-chair with Sanjay Singh as the member. It may be noted that Shah has also been made the chair of the ten-member committee constituted by the Government of India to draft the new National Water Policy of India. According to the information provided by the JJJA national convener, Sanjay Singh in January 2020, the draft of the water security bill was ready and has been shared with the public both within and outside the state for any recommendations or changes to make it better and effective. In its current form, the bill proposes, among other things, to:

- Recognise forest as the main source of flowing rivers in the state
- Make the forest department of the state work for maintaining and improving the forest cover
- Geo-tagging (digital photography or videography) of the public and private water resources
- Making strict provisions to put a tab on wastage of water
- Linking of farming with the monsoon patterns in the state
- Ensuring piped water supply in both urban and rural areas
- Strict implementation of pollution control board rules for water
- Strict compliance of rooftop rainwater harvesting system
- Creating awareness among masses for water conservation in the state by involving youths
- Regular meetings for water conservation at Panchayat levels
- Development of lakes at Panchayat levels

In addition to working towards bringing the water security law and becoming the first state to do so, the Government of Madhya Pradesh also decided to revive the old water conservation systems of the Bundelkhand region to make it drought-proof. It must be mentioned here that before the assembly elections in December 2018 in Madhya Pradesh, Parmarth had mobilised people in the rural areas of Bundlekhand. Parmarth officials, with the volunteers and community members, in large number, would go to the public rallies and addresses of the political leaders and would apprise them about the struggles for water in their areas. Activists of the campaign from the Bundelkhand region would hold dialogues with the candidates from different parties contesting elections and would seek assurances from them that after the formation of their government they would act towards addressing the issues of water in the region. And when the Indian National Congress formed the government in the state after the elections, they went to those representatives and prominent leaders reminding them of their promises made to the people of the Bundelkhand region.

During a symposium organised by JJJA in Bhopal in February 2019, two cabinet ministers from the Government of Madhya Pradesh participated along with many administrative officers. They assured that they would work towards addressing the water-related concerns of Bundelkhand soon. And later, they decided to identify 1000 old tanks from the Chandela and Bundela period and revive them by de-silting and repairing them besides clearing the channels that brought rainwater to them during the monsoon. The work of identifying these old water conservation structures was given to Parmarth after which their renovation would be taken up. Sanjay Singh informed that so far a thousand such tanks with an area of more than one hectare have been identified. Currently, a detailed project report (DPR) is being prepared by the government for starting the renovation work. He also informed that the work in this regard has slowed down a bit as the additional chief secretary (ACS) of the Panchayat and Rural Development Department, Gauri Singh was transferred in December 2019 who was personally taking a lot of interest in this work in Bundelkhand.

Talking about the other water-related talks with the government, Sanjay Singh informed that they had informed the government about the deteriorating condition of the rivers in the state and the government is focused on reviving the rivers. He said that at present, the Government of Madhya Pradesh has identified 40 rivers in the entire state and the rejuvenation work has started. He said,

I am happy that following the suggestions of the JJJA and Bundelkhand Water Forum (BWF – voluntary association formed in the region by the local civil society organisations and individuals), the state government has, in the first phase of the project, started the work on the rivers in Bundelkhand itself. Two rivers of Chhatarpur – Kel and Banne; Ur river of Tikamgarh, one in Damoh district and Mandakini river of Chitrakoot are being surveyed.

The district coordinator of Chhatarpur district, Ajay Yadav apprised that on Kel River the construction of conservation structures like check dams has also been started by the government in collaboration with the local communities. So, the advocacy work of JJJA has yielded good results, especially in the state of Madhya Pradesh, both in terms of pushing for the water security law and in protecting the rivers.

However, the Government of Madhya Pradesh was toppled with 22 MLAs of the Congress party resigning from the assembly in the month of March 2020 and the fate of the water security law in the state was still hanging in the balance. The national convener of the JJJA, who had informed in January 2020 that the state government was mulling about tabling the bill in the assembly during the monsoon session of 2020, said during a telephonic interview with one of the authors in July 2020 that "now, until the by-elections are held in the state, and a stable government is formed, the issue of the water security bill is not going to make any progress." However, he expressed hope that with the way the present chief minister of the state, Shivraj Singh Chouhan was associated with the work of JJJA during his previous governments until 2018, he will certainly take the bill forward and would not ignore it. He is committed to addressing the water-related issues in the state which he had also promised during the national water convention at Khajuraho in the Chhatarpur district of Madhya Pradesh in December 2020. "So, we are hopeful that we will be able to push the agenda strongly with this government too and current BJP-led government will not disappoint us either," Singh said.

Conclusion

Management of water, these days, is not just about creating new sources like dams, reservoirs, canals, pipelines, etc., but it is also about the judicious allocation of water for different competitive uses and at the same time maintaining the hydrological integrity of the existing water resources like lakes, rivers, their catchment areas, and the aquifers (Kumar, 2018). Integrity, here, refers to identifying and maintaining the natural characteristics of the resources. Kumar (2016) underlines that effective water management, in India, requires the understanding and application of ecological economics. Pointing to a similar concern, Mark Everard, an academic from the United Kingdom (UK) affiliated with the JJJA, during a Social and Environmental Justice Contemplation Camp at TBS Ashram, Alwar, reiterated that in India and many other countries, today, people are paying only for the infrastructure of the water supply like – reservoirs and pipelines. He said that when one water resource dries up, a pipeline is laid down to supply water from another resource which could be a reservoir or river for which cess is levied on the people. But none pays for the replenishment of the water resources and their catchment areas. Everard said that market economy has completely disconnected us from our natural world and its resources. He hailed the JJJA for its successful endeavours to make people fall in love with nature again.

The campaign, for instance, had mobilised people of ten villages in the Ghuwara block of the Chhatarpur district of Madhya Pradesh for rejuvenating a small river called Bachheri (around 10 kilometres long). Without any support from the government, the communities in these villages constructed

12 check dams on the river within three months. As a result of this, the river which had died was brought back to life after the monsoon of 2019. This reflects the potential of the civil society in terms of carrying out work in remote areas with local and marginalised communities as emphasised by Chandhoke (2007), Giri (2012), and Das (2007). The coming together of over 1200 organisations of different kinds from across the country for the restoration of the ailing water sector in India on the call of JJJA seems to be in sync with Shah (2019) who believes civil society organisations have a significant role to play in bringing people together and mobilising them for rights and advocate for favourable policies. The efforts of Parmarth and other civil society organisations associated through the campaign also seem to resonate with the idea of civil society organisations serving as "bridge-builders" between communities and other relevant stakeholders in working towards addressing contentious issues affecting people's lives, as noted by Das (2007, p. 47).

Similarly, the continuous demands of the JJJA before both the central and the state governments for bringing a water security law and devising policies for protecting water bodies appears analogous to what Khilnani (2002) says about civil society making demands before the state that it should accommodate and act for in the larger interest of the people. Civil society, the world over, has been considered as an important player, especially in terms of influencing the policies and investment (UNESCO, 2018). Mahajan's (1999) idea that in a democratic society, the government and the civil society must share a constructive relationship and work together for the promotion of citizenship rights could be observed when the Government of Madhya Pradesh partnered with the campaign leadership for drafting the water security bill to be tabled in the state assembly in 2019. The coming together of the Madhya Pradesh government and the JJJA re-establishes the idea of civil society and state being complementary to each as pointed out by Beteille (2001). This was evident also during the BJP-led government in the state in the year 2017 when the then chief minister of the state, Shivraj Singh Chouhan invited the JJJA members to be a part of the Narmada Seva Yatra and Namami Devi Narmade Yojana to conserve the Narmada River of the state and sought their advice for the project.

Thus, as pointed out by Edward (2004) that civil society is an outcome of the collective action of the people in search of good societies, JJJA is consistently making efforts throughout the country to protect the water bodies and to connect people to the ailing sector of water so that they could together come to its rescue. With several grassroots-level endeavours as a part of the JJJA making significant changes at the micro-level, the campaign, as a civil society initiative, also seems to possess an emancipatory potential for the people as outlined by Edward (2004). As various scholars consider civil society actors as important agents of change (Khilnani, 2002; Cohen & Arato, 1994; Mahajan, 1999), JJJA is bringing various stakeholders from various sections of the society together and facilitates the change-making processes, especially in the area of water management and protecting the water bodies which ultimately help to promote the socio-economic progress and at the same time also reverse the process of ecological degradation (Everard, 2015).

Notes

1 Taken from fieldnotes.
2 Parmarth Samaj Sevi Sansthan – Annual Report – 2016-2017.
3 Project Progress Report: Advocacy Efforts through Jal Jan Jodo Campaign; August-December 2014.
4 Ibid.
5 Parmarth Samaj Sevi Sansthan – Annual Report – 2016-2017.
6 Ibid.

6 Towards Strategic Water-Centred Environmental Communication

India, at present, is passing through a situation which is considered as extremely water-stressed and it is on the path to becoming a water-scarce country due to the widening gap between the demand and supply of this precious natural resource.[1] The river basins are becoming water-scarce and the groundwater levels are declining because of high exploitation of water resources by various sectors such as municipalities, industries agriculture, and others. This is also resulting in a number of water-related conflicts in the country.[2] This work is a study of the water management initiatives by a grassroots organisation, Parmarth, which works in the Bundelkhand region. Bundelkhand is a region that has experienced successive droughts in the last two decades and has witnessed several local conflicts owing to the dearth of water. The region is hot and dry, and droughts occur owing to drying water bodies triggered by the region experiencing deficient monsoon rainfall year after year.

At the time of the writing of the conclusion of the book, the newspapers in India were abuzz with the stories of around 100 women in the Bada Malhera block of Chhatarpur district of Madhya Pradesh who cut a mountain for 107 metres to make way for water to reach a 40-acre pond in the Agrotha village. This pond, developed under the Bundelkhand special package[3] in the year 2009–10, was meant to cater to the water needs of three villages of the block including – Agrotha, Bhelda, and Chaudharykheda. It took the women from the three villages about 18 months to cut the mountain and create a canal-like structure to bring the water to the pond. The work was carried out with the permission of the Forest Department of Madhya Pradesh. Parmarth played the role of a catalyst in facilitating the dialogue between the women and the officials of the Forest department. The officials of the department visited the site, interacted with the women and agreed to their idea, and approved the cutting of the forest. According to the district coordinator of Parmarth for Chhatarpur, many of the women were those who had returned to their villages following the nationwide lockdown imposed in March to contain the spread of COVID-19. He said that their involvement in the project expedited the work and the digging of the channel for the flow of rainwater to reach the pond from the mountains was completed in the month of August.

DOI: 10.4324/9781003350286-6

The women also decided to safeguard the area from the sand mining mafias as well as protect the trees of the area. They also met with the deputy forest ranger, Mahendra Singh Gour, and sought permission to protect every single tree in the area with a view to conserve nature and water. In addition, they also planted hundreds of trees along the newly created water channel so that the trees will hold the soil around it when they grow up and, thus, prevent erosion and siltation in it.

As mentioned in the book earlier, besides the local-level interventions in water management in association with the community members in the Bundelkhand region, Parmarth has also been running a nationwide campaign called *Jal Jan Jodo Abhiyan* (JJJA) (translated as water-people alliance campaign). This campaign aims to bring people working in different roles in the water and environmental sector together and advocate for better policies and actions by the government and the society to protect the water bodies in the country. It is also working towards pushing for a water security law on the lines of food security law in the country to ensure the availability of water for all.

Keeping this grand vision of Parmarth in focus, the conclusion chapter of the study has five sections. The first one discusses the important aspects from all the chapters in the book. As the study is based on an immersive critical analysis of the water management interventions being carried out by Parmarth in the Bundelkhand region and the JJJA campaign, this section also reflects on the learnings as derived from each of these thematic chapters. The following three sections endeavour to fathom deeper and convey the larger significance of the study by introducing expanded analytical ways of thinking about and further comprehending the field of environmental communication with a focus on water management issues. These sections build on the theoretical insights of this research study and go beyond to offer newer approaches for contextualising the research problem by dwelling on – (a) water communication; (b) communication for sustainable development; and (c) the relevance of Gandhian philosophies in contemporary environmental and water-related movements. The final section tries to tie the entire work together through some brief observations.

Key Reflections

This research was carried out with the objective of studying the water management interventions in India from the lens of environmental communication and communication for development (C4D) while also drawing from the concept of civil society. The goal was to unearth the communication ecologies of the environmental initiatives, especially those related to water conservation which so far appears to be a largely unexplored area in the Indian context. The chapters of the book as summarised below provide an overview of the content that it presents and a summary of the reflections in each of them.

The first chapter of the book, i.e. the "Introduction," outlines the water-related concerns around the world as well as in India, their evident causes, and reasons these issues need immediate and serious attention. The chapter also underlines some of the significant water management initiatives being undertaken by organisations in various parts of the world as well as India. It also provides the background for carrying out this research by examining the gaps in the field of environmental communication. It does so especially from the perspective of water-related challenges and the efforts being made by civil society groups to address these challenges in the Indian context. It also presents a theoretical background that later informs the research and helps in analysing the data collected from the field.

The second chapter gives details of the methodology adopted for this research, the fieldwork undertaken, and the methods used for data collection and analysis. The data for this research was collected through extensive and immersive fieldwork carried out in different parts of the Bundelkhand region of Uttar Pradesh and Madhya Pradesh. People's responses were gathered through formal and informal interactions to understand their everyday lived experiences, observations, struggles, etc. in the region during the fieldwork. Besides, the researchers also did participant-observation in various local events organised by Parmarth in Bundelkhand as well as in other parts of the country under its JJJA campaign. The data collection methods used by the researchers included in-depth interviews of the various stakeholders of the water management interventions at the local level in Bundelkhand as well as those participating in the *Jal Jan Jodo Abhiyan*, focus group discussion, document analysis, and participant-observation of the events.

In addition, the researchers also used methods such as "go-alongs" for empirical research and gathered substantive data and insights through observations, experiences, and opinions of respondents. The respondents were either those who were engaged in the micro-level interventions in various pockets of Bundelkhand or those working for the nationwide campaign of Parmarth. Such informal interactions, on the side-lines of different events, proved useful in obtaining a better understanding of the work related to water management due to the free-flowing nature of the conversations wherein the respondents do not have to make efforts to be articulate and conscious as in the case of formal interviews and FGDs as they are recorded.

The analysis of the data, available in the form of transcripts (of the interviews and focus group discussions), field notes, and analytic memos, was carried out through reading and coding to come up with concepts and ideas that emerged from them. The process of coding helped in linking the codes and categories with the theoretical framework of the research and segregating the data to shape the thematic chapters. Referring to the conceptual and theoretical frameworks perused as a part of review of literature also helped in analysing the data gathered from the field.

Using the conceptual framework derived from the "theory of change" approach, the third chapter explains the pathways determined by Parmarth for bringing about the change that it envisioned in terms of addressing the water-related issues of the villages. The chapter then maps the road traversed by Parmarth to make the change at the grassroots level, a reality, with support from the local communities, especially women in the rural areas. The "theory of change" chapter also analyses the stories of change collected from the field. The chapter is guided by the understanding derived from the readings of Jame (2013), Reeler and Blerk (2017), James (2013), Vogel (2012), Allen et al. (2017), and Anderson (2005). The central idea that emerges from the work of all these scholars defines theory of change as a process that helps an organisation decide the sequence of events that would lead to the desired outcome. Drawing from this literature, the authors trace the change-making process of Parmarth that has transformed some of the villages of Bundelkhand region from water-scarce to water-sufficient, viz. Moto and Kakradi (in Lalitpur district of Uttar Pradesh), Manpura (in Jhansi district), Itaha Kalpi (in Jalaun district of Uttar Pradesh), and others.

As the chapter delves deeper into the process of developing insights from the data to explain the nature of interventions for strengthening partnerships and developing communication strategies to see the change happen, it also draws from the scholars working on social norms from the perspective of social change. As Parmarth is working towards ensuring water security for the marginalised communities, which include the people from the lower castes and women, it also had to work towards changing the social norms in the villages where the deep-rooted patriarchal mindsets and attitudes were at the helm of determining the norms of everyday lives of the communities. So, to theorise the underlying approach in Parmarth's interventions the authors also referred to the work of scholars like Bicchieri (2006), Yanovitzky and Rimal (2006), Chung and Rimal (2016), Tankard and Paluck (2016), Bicchieri (2017), Bicchieri and Noah (2017), Vaitla et al. (2017), and others. Thus, through the "theory of change" framework, the book chronicles and explains how Parmarth starts its interventions by analysing the existing situation and understanding the social norms. It examines how the NGO then goes on to make women, important stakeholders, through persistent mobilisation efforts, in the initiatives to revive the old and dying water bodies in their villages and creating new water resources. It forms a collective of water at village level called *Pani Panchayat* which mostly has women as its members and the cadre of women that leads these collectives are identified as *Jal Sahelis* (female friends of water).

Initially, Parmarth had tried to mobilise both men and women together to work for addressing the water shortages in the rural areas but later after realising that women did not express themselves freely in the presence of men, they decided to keep the *Pani Panchayat* collectives as groups of women with only a few members as men. So, in a group of 20-member, *Pani Panchayat* would have 16–18 women and 2–4 men. After coming together through these groups and realising that they will themselves have to take control of the situation, the women started to take initiatives on their own for the renovation/repair of the existing water resources in their villages. They also started approaching the government authorities at different levels to apprise them about the issues they faced and sought their support for redressal. The *Pani Panchayat* groups, thus, also provided the women of the villages a kind of forum where they not only deliberated and discussed about the issues related to water but also other significant matters pertaining to their lives like sanitation, hygiene, employment, etc. and then communicate about them to the relevant authorities. Imarti Devi, a *Jal Saheli* of the Moto village of the Lalitpur district in Uttar Pradesh, during one of the FGDs had said, "*Pani Panchayat* groups have become a space where we can discuss all big and small issues that affect us and then we make our strategies to find solutions to them."

The fourth chapter draws its analytical basis from the conceptual framework of environmental communication and focuses on the communicative actions, practices, and strategies that underlie water management interventions. It analyses the communication strategies developed for mobilisation of the local communities, other stakeholders like community-based and voluntary organisations, significant individuals, besides, sensitising the local government authorities and public representatives to garner their support. The chapter cites the work of environmental communication scholar Anders Hansen (2015) who considers communication as playing a crucial role in constructing environmental problems as issues of public concern and also refers to Pezzulo and Cox (2018) who propagate environmental communication as a tool to understand the human relationship with the natural world. Through these and more readings, the study pieces together how the concept of environmental communication plays out in the case of Parmarth's initiatives to address the environmental problems by focusing primarily on the issues of water. The book argues that as suggested by Hansen (2015), Parmarth employs localised ways of communication – like engaging through dialogic communication, participating in the local festivals and gatherings of the communities, and using other participatory rural appraisal (PRA) tools, to discuss, understand, and construct the water-related problems as an issue of public concern. Through the work of scholars like Yang and Calhoun (2007) and Cox (2012), the study underlines how Parmarth and its engagement with other actors working in the water management sector results in creation of an "environmental" or "green" public sphere when they discuss and debate about the issues related to water and environment and take them forward.

Having examined the role of communication strategies and practices of Parmarth in its water management initiatives in various pockets of the Bundelkhand region, the fifth chapter then studies the activism and the advocacy efforts of Parmarth for a water security law through a civil society-led nationwide campaign called *Jal Jan Jodo Abhiyan* (JJJA). In addition to pressing for a water security law, the campaign also works for bringing people together from various sections of the society to protect the water bodies, especially rivers, many of which are drying up fast. This study finds out that besides identifying and mobilising local actors and stakeholders who play a constructive role in addressing the issues of water in the Bundelkhand region, Parmarth also gets in touch with the organisations and individuals working in the water management sector in various parts of the country and appeals to them to join hands to make the issue of water security an important national concern that demands immediate and effective action. The study draws from the scholars like Khilnani (2002), Cohen and Arato (1994), and Mahajan (1999) who consider civil society groups as important agents of change that touch people's lives in various ways. The chapter is also informed by Vinod (2006) and Beteille (2001) who assert that civil society plays a role in ensuring justice and equity for the marginalised sections of the society and in promoting citizenship and democracy. For analysing the JJJA campaign and the advocacy efforts spearheaded by Parmarth through motivating various civil society actors to push for a law ensuring water security, the study borrows from Bouman-Dentener and Devos (2015) who articulate civil society as an arena where people come together to advance a common interest that they care enough about. This "care" is clearly reflected in the work of Parmarth as a part of the campaign which has brought together hundreds of other organisations, groups, and individuals from various parts of the country who are collectively acting as activists to push the agenda of protecting the water bodies across the country by advocating the promulgation of the water security law.

The book also refers to the work of Khilnani (2002) who identifies civil society as an entity that makes demands before the state for better governability. In the case of the JJJA, on the one hand, both the central and the state governments seek inputs from the campaign leaders on the issues of water, and on the other, JJJA also regularly raises demands pertaining to the actions required by the governments to protect water bodies and ensure water for all. Sometimes, the "constructive relationship," as pointed out by Mahajan (1999), between the state and the civil society could also be gauged, especially when the governments invite the JJJA leaders to collaborate on several initiatives being undertaken by the governments to address the issues of water in various regions.

In all the initiatives discussed extensively in the thematic chapters, what was unique about the Parmarth's work and communicative actions was their effort to create a decentralised network of stakeholders who could contribute to the causes of water and other natural resources management activities in effective ways. In all its interventions, Parmarth gave a lot of importance to the local communities, self-help groups (SHGs), and community-based organisations (CBOs) that served as the pillars of support for all its initiatives.

These CBOs play a vital role in creating a strong link between Parmarth and the communities. Many times, when Parmarth organises events and invites government officials to interact with the communities, the members of the CBOs apprise them with the work they do and the issues they face related to water and other needs of life. So, once they are functional at their full potential, the CBOs serve as a critical link between Parmarth, communities, and the government authorities working in their areas. Parmarth keeps organising various kinds of capacity-building training programmes for them so that they could engage with concerned stakeholders at all levels and carry out the work in their areas with their support.

The research work attempts to understand the concerns and discourses related to water in the world as well as in India and, then, presents the interpretation of the stories of the water management initiatives at the grassroots level using the scholarship from the field of environmental communication and communication for development. Building upon the theoretical understanding of civil society's work within and outside India, it also explores and analyses the advocacy efforts around water security by various social actors. In doing so, the book also advances the possibilities of weaving together the thematic areas from this research and points to some emerging thrust areas within the field of environment and communication which may be explored further as the topics of academic enquiry. The three significant areas that emerge out of the discussions and analyses in the chapters of this book are: "water communication," "communication for sustainable development," and "the resonance of Gandhian philosophy in contemporary environmental/water movements."

Water Communication

Efforts related to water management and conservation are being made in various parts of the country in numerous ways depending on the nature of the water-related problems in specific areas. Most of the initiatives related to water management are being carried out with the support from the local communities. All kinds of change-making processes at the ground level, be it in the water sector or others, require community support which could be achieved only through an explicit communication strategy which will convey the benefits of a project to the communities (Ostrom, 2005). However, in most of the studies which have been carried out in the Indian context on various kinds of water-related projects, the dimension of "communication" does not surface as prominently. One would hardly come across a study carried out on a water management initiative that brings the communicative processes associated with them to fore. According to Ostrom (2005), another important function of communication is to facilitate discourses among the communities to develop a shared understanding of a problem and then ideating ways to deal with those problems. And this role of communication was evident from the various communicative activities that were carried out by Parmarth as a part of its local interventions as well as rallies and *yatras* of various kinds that helped people come together, participate in discussions about their water-related concerns, reach a common understanding, and devise mechanisms to find solutions to those issues.

Today, water crisis is becoming a global phenomenon and hence the efforts towards dealing with the crisis are also being made at various levels ranging from the international organisations to the local NGOs and CBOs. And the communication strategies and processes are integral parts of all such endeavours in the water management sector. Thus, for effective advocacy work, "water communication" needs to be promoted as a specialisation within the field of environmental communication for exclusive focus on water-related issues and communication strategies rooted in the understanding of the sector. This research draws heavily from the field of environmental communication, and while applying that understanding to water management, the authors felt that the water issue calls for a deeper and focused "water communication" to be taken up as a dedicated effort within the field. As emphasised by Herve-Bazin (2014), "water communication" applies to all kinds of communicative actions on fresh water resources of the world, including processes, interactions, messages, discourses, campaigns, etc. on water resources. She recognises communication as the key element for all kinds of planning and implementation of water management-related initiatives and identifies five key areas of research in water communication: "(1) Social Representation of Water; (2) Water Discourses; (3) Knowledge Brokerage; (4) Water Campaigns; and (5) Water Journalism" (p. XXIII).

From the work of Parmarth as well as of other organisations and individuals working on the issues of water, one can find the prevalence of all these areas highlighted by Herve-Bazin (2014). This study touches upon most of these concepts as it discusses the communication strategies, approaches, activities, social norms, etc. in the context of water management interventions by Parmarth. However, it also lays the foundation for digging deep into these specific areas for better visibility and integration of the communicative processes and approaches and situating them under the proposed domain of water communication. Although this study is Bundelkhand-centric, some phenomenal work is happening in terms of water management in various other parts of the country, including Rajasthan, Bihar, Uttarakhand, Madhya Pradesh, Maharashtra, Karnataka, Telangana, and others. The nature of interventions may vary depending on the topic focus of the projects, viz. water and sanitation, safe drinking water supply, water rights, natural water resource management, water supply for irrigation, rainwater harvesting, etc. However, all such programmes have elements of communication integrated into them in various forms like mobilisation, protests, negotiations with various stakeholders, advocacy, debates, discussions, etc.

Today, the entire world is concerned about the issues of environment and sustainability. And as various studies suggest, among the prominent topics related to environment and sustainability that are relevant to be studied from the perspective of communication and engagement are the concerns related to water resources (McCarthy, 2016; European Commission, 2014; Brown et al., 2013). Considering the amount of attention the area of water management is attracting from various quarters of the society, and the scale of work happening in the sector under the broad spectrum of environment and sustainability, the study puts a case for moving beyond the field of environmental communication and exploring the more nuanced area within the field, i.e. of water communication.

Kunz et al. (2017), in their study about communication gaps in water management sector, talk about some kinds of communication "silos" that exist between the various stakeholders working on water. Such "silos" exist in the water management sector in India as well. This was particularly underlined by the national convener of JJJA, Sanjay Singh while responding to the questions by the researchers on how did the idea of the campaign come to Parmarth? He said that when they saw people working on the issues of water and talking about them only with the stakeholders within their states and districts, "we realised that a lot of work is happening in various parts of the country but they were happening in isolation." Singh said that Parmarth discussed the idea of bridging the "communication gap" that existed among the individuals and organisations working on water in various parts of the country in numerous ways with the European Union which was supporting some of its local water management work in Bundelkhand region of India. That's how the idea of launching the *Jal Jan Jodo Abhiyan* (JJJA) was developed. So, creating a communication network of people and institutions working in the sector of natural resource management in general and water in particular was at the helm of this campaign. Singh reiterated that one of the major achievements of the campaign is that it is keeping all the relevant partners from all over the country engaged via events aimed at communicating on a regular basis within the network.

Herve-Bazin (2014) and Borca et al. (2016) advocate that communication is vital for all those involved in water management projects including policy-makers, environmental initiative groups, NGOs, regional authorities, and others to effectively take the messages related to water to the target population aimed at bringing some attitudinal change and driving actions. The water campaigns being carried out by various groups and institutions create various kinds of references that may influence the attitudes and behaviours of individuals as well as communities as experienced in the case of JJJA. And studies on these campaigns in terms of their significance in creating awareness, providing information, enabling communities to identify with the water-related challenges, facilitating dialogues, creating knowledge, and realising a shared responsibility to act and change the status quo could well be positioned under the domain of the "water communication" Herve-Bazin (2014).

Access to clean water and sanitation is one of the Sustainable Development Goals (SDGs) of the United Nations. And without ensuring the attainment of this goal, it would be extremely challenging to realise other goals of sustainable developments. According to United Nations (2018), SDG 6 (Clean Water and Sanitation) is crucial in making progress on all other SDGs as "sustainable management of water and sanitation underpins wider efforts to end poverty, advance sustainable development and sustain peace and stability" (p. 21). This report by United Nations (2018) further adds that the targets under the SDG 6 must be adapted in localised ways to the country's context. Many international development sector organisations that are working towards realising the different goals of sustainability as well as the governments realise the need for localising their interventions. That is where the local NGOs like Parmarth play the role of a catalyst between the governments, international agencies, and the local communities in terms of promoting dialogues and engagement between them. Effective engagement for any kind of development interventions requires good communication strategies to listen, respond, and be accountable to the needs of the people (Panos, 2007). The understanding of social norms and community attitudes is of paramount importance in any kind of change-seeking initiative, as such processes help in easing the implementation of the interventions and enhancing the outcomes desired at the beginning of a project (Syme & Hatfield-Dodds, 2007). This applies to the efforts being made to attain the sustainable development goals too. Therefore, it is crucial to understand the centrality of the role of communication in achieving sustainable development goals within the Indian context.

Communication for Sustainable Development in India

People are at the heart of any change-making process that seeks to attain a development-related goal. One of the best practices, when it comes to achieving sustainable development goals, as suggested by GTZ (2006) is involving people. And the other practices must include addressing the local needs, managing initiatives, forming alliances to share responsibility, and bringing sustainable development to the mainstream (GTZ, 2006). And communication has to play the central role in bringing various development actors together particularly in building the capacities of the people from the marginalised sections so that they can participate in the political as well as the development processes in effective ways (Panos, 2007).

When it comes to sustainable development, people from all walks of life must play a role in achieving its goals. And to bring people together to act, there is a need for a holistic communication approach that could embrace all sections of the societies, especially the poor and marginalised as well as all forms of communication, i.e. mediated or face-to-face. As the "Waterman of India," Rajendra Singh says,

> In India, there are a lot of good initiatives planned by the governments for achieving the sustainable development goals but they are all lying in the files. For the execution of those plans, governments need to make efforts to partner with the people in the society who have the actual potential to bring about the necessary changes to move towards sustainability by becoming important stakeholders in the implementation. Unfortunately, the governments are not able to do this except for a few individuals working in government institutions who are concerned about the environmental problems at personal level.

Singh further reiterated that until the government and the development agencies focus on communicative strategies and actions to involve various actors including the local communities in the sustainable development efforts, the Agenda 2030 could never be achieved. For him, most of the projects related to sustainable development and environmental protection fail in the country as the government and other responsible agencies do not make efforts to communicate the mission to the local people. According to United Nations (2016) the major challenge that lies in the implementation of the SDGs is taking the goals and targets of the agenda closer to the general public and telling them the benefits of achieving them and how they could contribute towards them in their everyday life.

There is a need for "whole-of-society approach" and "collaborative responses" for dealing with the complexities involved in the effective implementation of the SDGs (United Nations, 2016, p. 23). The report further emphasised on well-coordinated as well as closer cooperation between the governments and other stakeholders' groups so that a "broad ownership" of the SDGs could be created (United Nations, 2016, p. 23). The task of implementing the SDGs could not be done by only one organisation in isolation and needs multi-stakeholder engagement. It requires a variety of outreach activities and strategies for different types of actors and involvement of civil society for facilitating working interactions and open dialogues between the various stakeholders (United Nations, 2016). The secretary of Parmarth and national convener of JJJA, Sanjay Singh says,

> Today, when the entire world is talking about SDGs and Agenda 2030 and governments all over the world are prioritising the issue of sustainability of the earth's resources, the holistic and integrated approach that it requires appears to be missing. In a country like India, the most fundamental way to attaining the Agenda 2030 should be to focus on ensuring the sustainability of the local natural resources in smaller units like villages and Panchayats. And this requires sound thinking in terms of devising communication strategies and messages. Awareness-raising programmes like face-to-face dialogues, meetings, workshops etc. are

> to be planned for the communities which, in turn, also bring useful insights and inputs from those being made aware on the concerns of the sustainability. And when the communities understand and relate with the benefits they will get from a project, they become important partners in implementing them as well as the sustainability of the projects themselves.

There is a need to engage differently with different stakeholders by taking into account the realities of their lives and socio-economic conditions. Talking about the strategies of Parmarth and JJJA to communicate effectively with different actors, he says,

> We cannot go and talk to the rural communities about SDGs and Agenda 2030. We have to talk to them in the languages and vocabulary that they can relate with. Similarly when we have to meet the government officials and politicians, we have to talk in a different language and the message content changes completely. So, it's important for any organisation or group engaged in a project of sustainability to know the outcome they desire from the various interactions that they do with multiple stakeholders and accordingly choose the forms of communication as well as the message to be conveyed. We have seen hundreds of such projects fail due to the neglect of the aspect of communication process in the implementation of the work. It is pertinent to make different stakeholders realise their part of the responsibilities so that they can own them and contribute.

This research has underlined in detail the communication processes that take place at the grassroots level in the context of water management, and the researchers believe that for sustainable development, communication with people in the rural hinterlands of the country is indispensable, so that they, as Anne (1999) expresses, share their energy, talent, intellect, and commitment in mobilising each other and creating common resources for future usage. Mahatma Gandhi who is considered one of the greatest environmental thinkers in the world history (Bilimoria, 2001) talked about the issue of sustainability a century ago and emphasised on the need for judicious consumption of natural resources. The resonance of Gandhian thoughts and philosophies about nature and its resources could be seen at the core of the environmental initiatives and movements in the country. Taking into account the kind of relevance that Gandhi's ecological philosophies hold to our times, the researchers found scope in studying Gandhi as an environmental communicator, as well as adopting his methods in the communicational approaches ingrained in the endeavours aimed at sustainable development and environmental conservation.

Locating Gandhian Philosophy in Contemporary Environmental Movements

Environmental movements have been one of the important areas of study in the broader discipline of communication. India has seen many such movements like Chipko movement – 1973 in Uttarakhand, the Silent Valley movement – 1978 in Kerala, Jungle Bachao Andolan – 1982 in Bihar, Appiko Movement – 1983 in Karnataka, Narmada Bachao Andolan – 1985 in Gujarat, Madhya Pradesh and Maharashtra, Navdanya – 1987 in Uttarakhand and Karnataka, Tehri Dam Movement – 1990s in Uttarakhand, and others. Many of the environmentalists leading these movements include Medha Patkar, Sundarlal Bahuguna, Chandi Prasad Bhatt, Gaura Devi, and Vandana Shiva. Recently, Gyan Swaroop Sanand sacrificed his life for the cause of Ganga after observing a fast for 111 days. All of these leaders and activists are considered as Gandhian for their non-violent and peaceful mode of protests to press for their demands (Nandy, 2019; Guha, 1995; Kaushik, 2018).

According to Guha (1995), after the success of the Chipko movement to protect the forests of the Himalayas, a journalist announced that the trees in the Himalayan region have been saved by the ghost of Gandhi. Guha (1995) says that the prominent environmental activists like Bhatt, Bahuguna, Baba Amte, Patkar, Vandana Shiva, and others have heavily relied on Gandhian ways of non-violent protests or Satyagraha and have often expressed their own debt to Gandhi. These only prove that even as Gandhi is no more present physically in the world, his principles continue to resonate with many concerned with the well-being of the human and non-human world. The elements of Gandhian philosophies could be gauged in several small and big environmental movements in various parts of the country. If one looks at the work of Parmarth and JJJA, one would find that the kind of activism and other communicative actions that they engage with are non-violent and peaceful like doing rallies, river marches, Jal Satyagraha, dialogues, meetings, *Nadi Sansads* (river parliaments), and observing fast for pressurising governments to take actions against the degrading environmental conditions or the water bodies around them. This way, like other environmental movements, Parmarth also borrows heavily from the Gandhian ways of communicative actions for its local interventions in the Bundelkhand region as well as for the activities carried out under the JJJA.

Of late, the leaders and members of the JJJA have been continuously extending support and solidarity to Matri Sadan, a spiritual Ashram (centre) located at Haridwar in Uttarakhand, where spiritual persons have been continuously observing fasts one after another for demanding action from the central government to protect the River Ganga and stop the proposed hydroelectric projects on the holy river. Swami Nigamananda and Swami Gyan Swaroop Sanand (a former IIT professor) had sacrificed their lives in the year 2011 and 2018 respectively while observing fast for over 100 days demanding action from the government to protect the national river from deterioration due to the proposed dams and hydro projects on the river. Following Swami Gyan Swaroop's demise in October 2018, the other monks at the Ashram had continuously been fasting for the cause. Among those who picked up the baton since 2018 included Sadhvi Padmavati and Brahmachari Atmabodhanand both of whom had observed fast for more than 100 days. Swami Shivananda has been the recent one to go on fast (only consuming four glasses of water daily) who started his agitation on August 3, 2020. However, after a month of fast he received a communication from the Union Jal Shakti Minister Gajendra Singh Shekhawat and director of National Water Mission for Clean Ganga (NMCG) Rajeev Ranjan Mishra assuring to look into his demands for protecting the River Ganga, following which Swami Shivananda broke his fast. The JJJA leader like Rajendra Singh, Sanjay Singh, and others have continuously been visiting the Ashram for expressing their solidarity with the seers agitating for the cause of the River Ganga and participating in the dialogues with the monks and the government officials for a breakthrough in this struggle. Resorting to fasting as a non-violent and peaceful means of protests was an act mastered by Mahatma Gandhi (Mukherjee, 2018) that inspired many leaders of India's freedom movement as well as other parts of the world for peaceful resistance against the power structures. After Gandhi, the non-violent ways of resistance become a norm for peace-loving leaders and activists world over. And his communicative actions like *padyatra* (foot-march), face-to-face dialogues, hunger strikes, etc. were extensively embraced by the environmental movements and groups in India.

Resemblances of Gandhian ideals could be gauged even in the grassroots-level initiatives on environment and water in India. A *Jal Saheli* of the Moto Village of Lalitpur district in Bundelkhand was once asked as to what they communicate to the women in their and the nearby villages to include them in the collective called *Pani Panchayat*? She said,

> We try to tell them the benefits (for the women community) of the work that we do as a group and give them the examples of the such benefits that other groups' members have reaped by coming together. Sometimes, women and their family members understand easily and they become one of us but sometimes it does not work that way. In that case, we simply focus on our work and believe that when they see what we can do as women groups they will certainly join hands with us. So, sometimes we let our actions speak to them instead of the words.

This resonates with what Gandhi wrote to the organisers of World Fellowship Faiths on October 2, 1933, when they asked the Mahatma to send a birthday message, "What message can I send through the pen, if I am not sending any through the life I am living?"[4]

In Conclusion

This study has examined the water management interventions at grassroots level in the Bundelkhand region of India led by a local NGO in addition to examining a nationwide campaign on water run by it called *Jal Jan Jodo Abhiyan* (JJJA). This research has been carried out under the broad themes of environmental communication, civil society, and theory of change using qualitative research methods and tools. Although it is a study carried out by taking up an NGO as a case study, the stories and insights that it provides from the field strongly connect with the thematic areas of environmental and communication, communication for development, and role of civil society in water management in India.

It also opens up possibilities for future research on the plethora of water management interventions being undertaken by the governments, civil society organisations, NGOs, CBOs, and individuals in different parts of the country from different perspectives in the field of communication. The authors consider that there is a need to talk in terms of "water communication" considering the crucial manner in which water affects both the social and the economic lives of the people and the nation. Water is becoming scarcer all over the world due to the erratic pattern of consumption and indiscriminate exploitation of natural resources which are putting tremendous stress on the freshwater bodies of the world. A field of study like "water communication" would aim at not only addressing the causes of the dying water bodies and people's struggle for water but also find out and communicate the success stories of effective water management initiatives all over the world that could be the way forward.

Notes

1 State of India's Environment – 2020 published by Centre for Science and Environment.
2 Ibid.
3 For more details, refer to:http://mpplanningcommission.gov.in/bundelkhand/bundelkhand.htm.
4 Anecdote of Gandhi compiled by RK Prabhu published in July 1954 by Navajivan Publishing House, Ahmedabad.

References

Abbati, M. (2019). *Communicating the environment to save the planet: A journey into Eco-communication*. Switzerland: Springer.

Ahmad, F. (1992). Popular people's movements in water resources management and the role of law. In C. Singh (Ed.), *Water law in India* (pp. 243–259). New Delhi: Indian Law Institute.

Ahmad, N. (2018). The role of civil society institutions in environmental governance in India: Post-colonial context and human rights challenges in the environmental justice. *International Journal of Legal Studies and Research*, *Special Issue*, 16–39.

Ahmed, A., Deb, D., & Mondal, S. (2019). Assessment of rainfall variability and its impact on groundnut yield in Bundelkhand region of India. *Current Science*, *117*(5), 794–803.

Allen, W., Cruz, J., & Warburton, B. (2017). How decision support systems can benefit from a theory of change approach. *Environmental Management*. The final publication is retrieved from Springer via: https://dx.doi.org/10.1007/s00267-017-0839-y.

Anderson, A. (2015). Reflections on environmental communication and the challenges of a new research agenda. *Environmental Communication*, *9*(3), 379–383.

Anderson, A. A. (2005). *The community builder's approach to theory of change: A practical guide to theory development*. New York: Aspen Institute Roundtable on Community Change.

Anderson, A. G. (2014). *Media, environment and the network society*. London: Palgrave Macmillan.

An Inter-Agency Initiative. (2017). *Assessment of impact of drought on men, women and children*. CASA, Action Aid, UNICEF, OXFAM, IAGUP, WaterAid.

Anne, S. S. (1999). Synergizing participation: Are you able to enable? In S. A. White (Ed.), *The art of facilitating participation: Releasing the power of grassroots communication* (pp. 69–79). New Delhi: Sage Publications.

Anyaegbunam, C., Mefalopulos, P., & Moetsabi, T. (1999). Facilitating grassroots participation in development: New training models and techniques. In S. A. White (Ed.), *The art of facilitating participation: Releasing the power of grassroots communication* (pp. 207–228). New Delhi: Sage Publications.

Baba, J. J. (2015, February 8). *Dominant paradigm of development*. Retrieved April 12, 2016, from Communication and Journalism: http://wecommunication.blogspot.in/2015/02/dominant-paradigm-of-development.html.

Barber, B. R. (1995). Searching for civil society. *National Civic Review*, *84* (2), 114–118.
Barbour, R. (2007). *Doing focus groups*. London: Sage Publications.
Benaquisto, L. (2008). Codes and coding. In L. M. Given (Ed.), *The Sage Encyclopedia of qualitative research methods* (Vol. 1 & 2, pp. 85–88). Thousand Oaks: Sage.
Benaquisto, L. (2008). Open coding. In L. M.Given (Ed.), *The Sage Encyclopedia of qualitative research methods* (Vol. 1 & 2, pp. 581–582). Thousand Oaks: Sage.
Beteille, A. (2001). Civil society and the good society. *Sociological Bulletin*, *50* (2), 286–307.
Bhatt, A. (1995). Voluntary action in India: Role, trends and challenges. *Economic and Political Weekly*, *30* (16), 870–873.
Bicchieri, C. (2017). *Norms in the wild: How to diagnose, measure, and change social norms*. New York: Oxford University Press.
Bicchieri, C. (2006). *The grammar of society: The nature and dynamics of social norms*. New York: Cambridge University Press.
Bicchieri, C., & Noah, T. (2017). Applying social norms theory in CATS programming. Retrieved November 10, 2018, from PENN SoNG: https://repository.upenn.edu/cgi/viewcontent.cgi?article=1014&context=pennsong.
Bigas, H. (Ed.). (2012). *The global water crisis: Addressing an urgent security issue*. Papers for the InterAction Council, 2011–2012. Hamilton, Canada: UNU-INWEH.
Bilimoria, P. (2001). Mahatma Gandhi. In J. A. Palmer (Ed.), *Fifty key thinkers on the environment* (pp. 160–166). London: Routledge.
Borca, C., Sarca, I., Mihartescu, A.-A., & Bogdan, P. (2016, May 25). *Best communication practices in the water supply sector*. Retrieved September 15, 2020, from To Know Press: http://www.toknowpress.net/ISBN/978-961-6914-16-1/papers/ML16-159.pdf.
Bouman-Dentener, A., & Devos, B. (2015). Civil Society: key contributors to water and sustainable development. *Water and Sustainable Development: From vision to action* (pp. 1-17). Zaragoza: UN Water.
Bowonder, B. (1987). Management of environment in developing countries. *The Environmentalist*, 7, 111–122.
Boykoff, M. T., & Roberts, J. T. (2007). *Media coverage of climate change: Current trends, strengths, weaknesses*. New York: UNDP.
Brown, E., Hartman, K., Borick, C. P., Rabe, B. G., & Ivacko, T. M. (2013). *The national surveys on energy and environment public opinion on fracking: Perspectives from Michigan and Pennsylvania*. Ann Arbor: Center for Local, State, and Urban Policy (CLOSUP).
Burke, K. (1966). *Language as symbolic action*. Berkeley: University of California Press.
Care Nepal. (2012). *Theories of change in peacebuilding: Learning from the experiences of peacebuilding in Nepal*. Khathmandu: Care Nepal.
Carpiano, R. M. (2009). Come take a walk with me: The "Go-Along" interview as a novel method for studying the implications of place for health and well-being. *Health & Place*, *15*(1), 263–272.
Carson, R. (1962). *Silent spring*. Boston: Houghton Mifflin.
Chandhoke, N. (2003). *The Conceits of Civil Society*. New Delhi: Oxford University Press.
Chandhoke, N. (2007). Civil society. *Development in Practice*, *17* (4–5), 607–614.

Chauhan, C. (2017, September 13). States stare at drought as 59% of India get deficit rainfall, food output at the lowest. Retrieved December 12, 2018, from Hindustan Times: https://www.hindustantimes.com/india-news/states-stare-at-drought-as-59-of-india-get-deficit-rainfall-food-output-at-the-lowest/story-L754PYi3N6HCoCVQ2lGXHJ.html.

Chung, A., & Rimal, R. N. (2016). Social norms: A review. *Review of Communication Research*, *4*(1), 1–28.

Cohen, J. L., & Arato, A. (1994). *Civil society and political theory*. Cambridge: MIT Press.

Corbett, J. B. (2006). *Communicating nature: How we create and understand environmental messages*. Washington, DC: Island Press.

Corbin, J., & Strauss, A. (2008). *Basics of qualitative research: Techniques and procedures for developing grounded theory*. Thousand Oaks: Sage.

Cox, R. (2007). Nature's 'crisis discipline': Does environmental communication have an ethical duty? *Environmental Communication*, *1*(1), 5–20.

Cox, R. (2012). *Environmental communication and the public sphere*. Thousand Oaks, CA: Sage Publications.

Cox, R., & Depoe, S. (2015). Emergence and growth of the "field" of environmental communication. In A. Hansen, & R. Cox (Eds.), *The Routledge Handbook of environment and communication* (pp. 13–25). Oxon: Routledge.

Crist, M. (2019, April 9). Down to earth: Why is the story of climate catastrophe so hard to tell? Retrieved May 2, 2019, from The New Republic: https://newrepublic.com/article/153505/case-climate-alarmism-nathaniel-rich-david-wallace-wells-review.

Darlington, Y., & Scott, D. (2002). *Qualitative research in practice: Stories from the field*. Sydney: Allen & Unwin.

Das, S. K. (2007). *Conflict and peace in India's northeast: The role of civil society*. Washington, DC: East-West Center Washington.

Dasgupta, K. (2016, May 10). *Don't blame nature for the drought in Bundelkhand*. Retrieved February 18, 2017, from Hindustan Times: http://www.hindustantimes.com/editorials/don-t-blame-nature-for-the-drought-in-bundelkhand/story-4ZJjFLdYJ3Tq8pN1sJvrsO.html.

Davies, R. (2004). Scale, complexity and the representation of theories of change. *Evaluation*, *10*(1), 101–121.

Development Alternatives. (2012). *Sustainable civil society initiative to address global environmental challenges*. New Delhi: Development Alternatives.

Development Alternatives. (2015). *Climate resilient development in Bundelkhand region of Madhya Pradesh: Vulnerability and adaptation assessment*. Development Alternatives.

Down To Earth. (2015, June 28). *Rise of third sector*. Retrieved February 20, 2020, from Down to Earth: https://www.downtoearth.org.in/coverage/rise-of-third-sector-33712.

Down To Earth. (2015, July 13). *Rivers, up close and personal*. Retrieved November 16, 2020, from Down To Earth: https://www.downtoearth.org.in/coverage/water/rivers-up-close-and-personal-41907.

Dreiling, M., Lougee, N., Jonna, R., & Nakamura, T. (2008). Environmental organisations and communication praxis: A study of communication strategies among a national sample of environmental organizations. *Organization & Environment*, *21* (4), 420–445.

EBTC. (2011). *Water management sector in India: An overview of research and activity*. New Delhi: European Business and Technology Centre.
Edwards, M. (2004). *Civil society*. Cambridge: Polity Press.
European Commission. (2014). *Special eurobarometer 419: Public perception of science, research and innovation.* Brussels: Directorate-General for Communication.
Everard, M. (2015). Community-based groundwater and ecosystem restoration in semi-arid north Rajasthan (1): Socio-economic progress and lessons for groundwater-dependent areas. *Ecosystem Services*, *16*(1), 125–135.
FAO. (2008). *Water and the rural poor: Interventions for improving livelihoods in sub-Saharan Africa*. Rome: Food and Agriculture Organisation.
Friedman, S. M. (2004). And the beat goes on: The third decade of environmental journalism. In S. L. Senecah (Ed.), *The environmental communication yearbook – volume 1* (pp. 175–188). London: LEA.
Gemmil, B., & Bamidele-Izu, A. (2002). The role of NGOs and civil society in global environmental governance. In D. C. Esty & M. H. Ivanova (Eds.), *Global environmental governance: Options & opportunities* (pp. 77–100). New Haven: Yale School of Forestry & Environmental Studies.
Gerring, J. (2007). *Case study research: Principles and practice.* Cambridge: Cambridge University Press.
Gillham, B. (2000). *Case study research methods.* London & New York: Continuum.
Giri, A. K. (2012). Rethinking civil society and the public sphere: Pathways of Indian modernities in global conversations. *Polish Sociological Review*, *178* (2), 227–243.
Gobo, G. (2008). *Doing ethnography* (A. Belton, Trans.). London: Sage.
GTZ. (2006). *Strategic communication for sustainable development: A conceptual overview*. Eschborn: GTZ.
Guha, R. (1995). Mahatma Gandhi and the environmental movement in India. *Capitalism Nature Socialism*, *6*(3), 47–61.
Guppy, L., & Anderson, K. (2017). *Water crisis report.* Hamilton: United Nations University Institute for Water, Environment and Health.
Hancock, D. R., & Algozzine, B. (2016). *Doing case study research.* New York: Teachers College Press.
Hansen, A. (2015). Communication, media and the social construction of the environment. In A. Hansen & R. Cox (Eds.), *The Routledge Handbook of environment and communication* (pp. 26–38). Oxon: Routledge.
Herve-Bazin, C. (2014). Scope of water communication. In C. Herve-Bazin (Ed.), *Water communication: Analysis of strategies and campaigns from water sector* (pp. 1–34). London: IWA Publishing.
Hussain, I. (2007). Pro-Poor intervention strategies in irrigated agriculture in Asia: Issues, lessons, options and guidelines. *Irrigation and Drainage*, *56*(2–3), 119–126.
IDSA. (2010). *Water security for India: The external dynamics*. New Delhi: Institute for Defence Studies and Analyses.
James, C. (2011). *Theory of change review. A report commissioned by comic relief.* London: Comic Relief.
James, C. (2013). *Theory of change – A guide for small and diaspora NGOs.* INTRAC (Remote organisation registered in England) . Retrieved from: https://

www.intrac.org/wpcms/wp-content/uploads/2016/09/Theory-of-change_A-Guide-for-Small-and-Diaspora-NGOs-1.pdf.

Jitendra. (2016, May 4). Penny unwise, pond foolish. Retrieved November 22, 2018, from Down To Earth: https://www.downtoearth.org.in/news/agriculture/it-is-not-a-drought-53812.

Joshi, H., & Kumar, A. (2007). Role of non-governmental organizations in restoration and management of freshwater ecosystems. *Proceedings of Taal 2007: The 12th World Lake Conference*, pp. 1880–1884. Jaipur: Ministry of Environment & Forest, Government of India.

Jurin, R., Roush, D., & Danter, J. (2010). *Environmental communication: Skills and principles for natural resource managers, scientists, and engineers*. New York: Springer.

Kamat, S. (2002). *Development hegemony: NGOs and the state in India*. New Delhi: Oxford University Press.

Kashyap, S. K. (2019, June 4). *55% in the availability of fresh water since 1960; 60 crore Indians experience severe water distress*. Retrieved November 1, 2020, from The Logical Indian: https://thelogicalindian.com/story-feed/awareness/water-crisis/.

Kaushik, R. (2018, July 4). *Lessons for the environmental movements from Gandhi*. Retrieved August 2, 2019, from countercurrent.org: https://countercurrents.org/2018/07/lessons-for-the-environmental-movement-from-gandhi.

Khilnani, S. (2002). The development of civil society. In S. Kaviraj & S. Khilnani (Eds.), *Civil society: History and possibilities* (pp. 11–32). New Delhi: Cambridge University Press.

Koul, D. N., Singh, S., Neelam, G., & Shukla, G. (2012). Traditional water management systems – An overview of Ahar-Pyne system in South Bihar plains of India and need for its revival. *Indian Journal of Traditional Knowledge*, *11*, 266–272.

Kumar, M. D. (2018). Proposing a solution to India's water crisis: 'paradigm shift' or pushing outdated concepts? *International Journal of Water Resources Development*, *34*, 42–50.

Kumar, R., Singh, R. D., & Sharma, K. D. (2005). Water resources in India. *Current Science*, *89* (5), 794–811.

Kunz, N. C., Kastelle, T., & Moran, C. J. (2017). Social network analysis reveals that communication gaps may prevent effective water management in the mining sector. *Journal of Cleaner Production*, *148*(1), 915–922.

Kusenbach, M. (2003). Street phenomenology: The go-along as ethnographic research tool. *Ethnography*, *4*(3), 455–485.

Leat, D. (2005). *Theories of social change*. Gutersloh: Bertelsman Stiftung.

Mahajan, G. (1999). Civil society and its avatars: What happened to freedom and democracy. *Economic and Political Weekly*, *34* (20), 1188–1196.

Mahapatra, R. (2017, October 26). *It is not a drought but a cumulative outcome of decades of policy sins*. Retrieved December 15, 2018, from Down To Earth: https://www.downtoearth.org.in/news/agriculture/it-is-not-a-drought-53812.

Manyozo, L. (2012). *Media, communication and development: Three approaches*. New Delhi: Sage Publications.

McCarthy, J. (2016, March 17). *American's concern about water pollution edge up*. Retrieved September 12, 2020, from Gallup: https://news.gallup.com/poll/190034/americans-concerns-water-pollution-edge.aspx.

McKechnie, L. E. (2008). Observational research. In L. M. Given (Ed.), *The Sage encyclopedia of qualitative research methods* (pp. 573–575). Thousand Oaks: Sage Publications.

McKee, N., Beckner-Benton, A., & Bockh, E. (2014). Social and behavior change communication. In K. G. Wilkins, T. Tufte, & R. Obregon (Eds.), *The Handbook of development communication and social change* (pp. 278–297). Hoboken, NJ: John Wiley & Sons.

Meisner, M. (2015, November). Environmental communication: What it is and why it matters. Retrieved May 22, 2019, from The IECA: https://theieca.org/resources/environmental-communication-what-it-and-why-it-matters.

Mihir Shah Committee. (2016). *A 21st century institutional architecture for India's water reforms*. New Delhi: MoWR, RD&GR.

Milsten, T. (2009). Environmental communication theories. In S. W. Foss (Ed.), *Encyclopedia of communication theory* (pp. 344–348). New Delhi: Sage Publications Inc.

Mishra, M. (2013). Environmental communication in India: Lessons from Orissa. *Media Asia*, *40*, 85–95.

Moemeka, A. A. (1994). Development communication: A historical and conceptual overview. In A. A. Moemeka (Ed.), *Communicating for development: A new pan-disciplinary perspective* (pp. 3–21). New York: State University of New York.

Morgan, D. L. (2008). Focus groups. In L. M. Given (Ed.), *The Sage encyclopedia of qualitative research methods* (pp. 352–354). Thousand Oaks: Sage Publications.

Morgan, E. L. (2003). Discourses of water: A framework for the study of environmental communication. *Applied Environmental Education and Communication: An International Journal*, 2(3), 153–159.

MoWR. (2011). *National water mission under national action plan on climate change*. New Delhi: Ministry of Water Resources, Government of India.

Mukherjee, S. (2018, April 11). *Fast: Powerful protest tool has a long history*. Retrieved September 25, 2020, from Deccan Chronicle: https://www.deccanchronicle.com/opinion/columnists/110418/fast-powerful-protest-tool-has-a-long-history.html.

Najam, A., Runnalls, D., & Halle, M. (2007). *Environment and globalization: Five propositions*. Manitoba: International Institute for Sustainable Development.

Nandy, A. (2019, January 30). Relevance of Gandhian environmentalism (K. D. Gupta, Interviewer). New Delhi: Down To Earth.

NCW. (2005). *Water and women*. New Delhi: National Commission for Women.

NIDM. (2014). *Bundelkhand drought: Retrospective analysis and way ahead*. New Delhi: National Institute of Disaster Management.

NITI Aayog. (2019). *Composite water management index*. New Delhi: Niti Aayog.

Nofri, S. (2011). *Cultures of environmental communication*. Hamburg: Springer VS.

OECD. (2012). *OECD environmental outlook to 2050: The consequences of inaction*. Paris: OECD Publishing.

Ostrom, E. (2005). *Understanding institutional diversity*. Princeton: Princeton University Press.

Pacific Institute. (2009). *Water scarcity & climate change: Growing risks for business & investors*. Oakland: Pacific Institute.

Panos. (2007). *The case for communication in sustainable development*. London: Panos London.

Parekh, B. (2004). Putting civil society in its place. In M. Glasius, D. Lewis, & H. Seckinelgin (Eds.), *Exploring civil society: Political and cultural contexts* (pp. 15–25). London: Routledge.

Parikh, K. S. (2013, August). *India's water crisis: Causes and cures*. NBR, Interviewer.

Perspectives. (2010). Drought by design: The Man-made calamity in Bundelkhand. *EPW*, *45*(5), 33–38.

Pezzulo, P. C., & Cox, R. (2018). *Environmental communication and the public sphere*. Los Angeles: Sage.

POMIO. (2014, November). *Guidelines for improving water use efficiency in irrigation, domestic & industrial sectors*. New Delhi: Ministry of Water Resources, Government of India.

Princen, T., & Finger, M. (1994). Introduction. In T. Princen & M. Finger (Eds.), *Environmental NGOs in world politics: Linking the local and the global* (pp. 1–28). London: Routledge.

Prior, L. F. (2008). Document analysis. In L. M. Given (Ed.), *Sage encyclopedia of qualitative research methods* (pp. 230–231). Thousand Oaks: Sage Publications.

Reeler, D., & Blerk, R. V. (2017). *The truth of the work: Theory of change in a changing world*. Cape Town: CDRA.

Retzinger, J. (2010). Spectacles of labor: Viewing food production through a television screen. *Environmental Communication*, *4*(4), 441–460.

Rogers, P. (2014). *Overview of impact evaluation*. Florence: UNICEF.

Salopek, P. (2019, August 9). *India's holiest river is drying up*. Retrieved November 10, 2020, from National Geographic: https://www.nationalgeographic.com/culture/2019/08/india-ganges-river-drying-up/.

Samuel, R., & Thanikachalam, V. (2003). Non-governmental organization (Ngos) spearheading public participation in environmental issues. In Martin J. Bunch, V. Madha Suresh, & T. Vasantha Kumaran (Eds.), *Proceedings of the Third International Conference on Environment and Health*, Chennai, India, 15–17 December, 2003, pp. 427–440.

SANDRP. (2018, October 12). *Bundelkhand: Overview of 2018 monsoon*. Retrieved November 14, 2020, from SANDRP: https://sandrp.in/2018/10/12/bundelkhand-overview-of-2018-monsoon/.

Sapru, S. (2019, July 10). *Time is running out*. Retrieved November 3, 2020, from GenderMatters.in: https://gendermatters.in/time-is-running-out/.

Schensul, J. J. (2008a). Documents. In L. M. Given (Ed.), *The Sage encyclopedia of qualitative research methods* (p. 232). Thousand Oaks: Sage Publications.

Schensul, J. J. (2008a). Methodology. In L. M. Given (Ed.), *The Sage encyclopedia of qualitative research methods* (Vol. 1 & 2, pp. 516–521). Thousand Oaks: Sage.

Schoenfeld, C. (1969). What's new about environmental education? *The Journal of Environmental Education*, *1*(1), 1–4.

Schramm, W. (1964). *Mass media and national development*. Stanford: Stanford University Press.

Servaes, J. (2008). Introduction. In J. Servaes (Ed.), *Communication for development and social change* (pp. 14–28). New Delhi: Sage Publication.

Servaes, J., & Malikhao, P. (2002). Development communication approaches in an international perspective. In J. Servaes (Ed.), *Approaches to development communication* (pp. 1–38). Paris: UNESCO.

Shah, G. (2019). *Democracy, civil society and governance*. New Delhi: Sage Publications.

Shah, T., & Koppen, B.V. (2016). The precept and practice of Integrated Water Resource Management (IWRM) in India. In V. Narain & A. Narayanamoorthy (Eds.), *Indian water policy at the crossroads: Resources, technology and reforms* (pp. 15–33). London: Springer.

Singh, J. (2014, June 9). *Bundelkhand's cycle of droughts: Is it man-made?* Retrieved February 15, 2017, from Down to Earth: http://www.downtoearth.org.in/news/bundelkhands-cycle-of-droughts-is-it-manmade-44677.

Singh, K. (2000). Managing India's water resources. In *Government and civil society partnerships* (pp. 15–16). Hyderabad: Swiss Centre for Development Cooperation in Technology and Management.

Singhal, A. (2001). *Facilitating community participation through communication*. New York: UNICEF.

Sowards, S. K. (2012). Introduction: Environmental justice in international contexts: Understanding intersections for social justice in the twenty-first century. *Environmental Communication*, *6*(3), 285–289.

Stern, P. C. (2000). Toward a coherent theory of environmentally significant behaviour. *Journal of Social Issues*, *56*(3), 407–424.

Syme, G. J., & Hatfield-Dodds, S. (2007). The role of communication and attitudes research in the evolution of effective resource management arrangements. In K. Hussey & S. Dovers (Eds.), *Managing water for Australia: The social and institutional changes* (pp. 11–21). Collingwood: CSIRO Publishing.

Tankard, M. E., & Paluck, E. L. (2016). Norm perception as a vehicle for social change. *Social Issues and Policy Review*, *10*(1), 181–211.

Tufte, T. (2017). *Communication and social change: A citizen perspective*. Cambridge: Polity Press.

UNDP. (2007). *Uttar Pradesh Human Development Report*. New York: United Nations Development Programme.

UNDP. (2019). *Human development report 2019*. New York: United Nations Development Programme.

UNESCO. (1997). *Educating for a sustainable future: A transdisciplinary vision for concerted action*. Paris: UNESCO.

UNESCO. (2012). *Managing water under uncertainty and risk: The United Nations world water development report 4, volume 1*. Paris: UNESCO.

UNESCO. (2018). *The United Nations world water development report*. Paris: UNESCO.

UNICEF. (2007). *Women and children: The double dividend of equality*. New York: UNICEF.

UNICEF, FAO & SaciWaters. (2013). *Water in India: Situation and prospects*. New Delhi: UNICEF.

United Nations. (2016). *Synthesis of voluntary national reviews*. Division for Sustainable Development & New York: Department of Economic and Social Affairs.

United Nations. (2018). *Sustainable development goal 6: Synthesis report on water and sanitation*. New York: United Nations.

Vaitla, B., Taylor, A., Horn, J. V., & Cislaghi, B. (2017). *Social norms and girls' well-being: Linking theory and practice*. Washington, DC: Data2X.

ven den Hoonard, D. K., & ven den Hoonard, W. C. (2008). Data analysis. In L. M. Given (Ed.), *The Sage Encyclopedia of qualitative research methods* (pp. 186–188). Thousand Oaks: Sage Publications.

Verschuren, P., & Doorewaard, H. (2010). *Designing a research project*. The Hague: Eleven International Publishing.

Vinod, M. (2006). The changing dimensions of civil society in the twenty-first century: Theory versus reality. *The Indian Journal of Political Science*, *67* (4), 783–792.

Vinod, M. J., & Deshpande, M. (2013). *Contemporary political theory*. Delhi: PHI Learning Private Limited.

Vogel, I. (2012). *Review of the use of 'theory of change' in international development*. London: UK Review Report, Department for International Development.

Water Aid. (2008). *Drought and drinking water crisis in Bundelkhand: Half full half empty*. New Delhi: Water Aid.

White, S. A. (Ed.). (1999). *The art of facilitating participation: Releasing the power of grassroots communication*. New Delhi: Sage Publications.

Wishart, G. (2000). *Government and civil society partnerships*. St. Gallen: HTN-SKAT.

Wood, G. (1997). States without citizens: The problem of the franchise state. In D. Hulme & M. Edwards (Eds.), *NGOs, states, and donors: Too close for comfort?* (pp. 79–93). London: Macmillan Press.

World Bank. (2017, September 20). *Water resource management*. Retrieved October 28, 2020, from The World Bank: https://www.worldbank.org/en/topic/waterresourcesmanagement#1.

Yang, G., & Calhoun, C. (2007). Media, civil society, and the rise of a green public sphere in China. *China Information*, *21* (2), 211–236.

Yanovitzky, I., & Rimal, R. N. (2006). Communication and normative influence: An introduction to the special issue. *Communication Theory*, *16*, 1–6.

Yin, R. K. (2003). *Case study research: Design and methods*. Thousand Oaks: Sage Publication.

Yin, R. K. (2011). *Qualitative research from start to finish*. New York: The Guildford Press.

Index

For Product Safety Concerns and Information please contact our EU
representative GPSR@taylorandfrancis.com
Taylor & Francis Verlag GmbH, Kaufingerstraße 24, 80331 München, Germany

www.ingramcontent.com/pod-product-compliance
Ingram Content Group UK Ltd.
Pitfield, Milton Keynes, MK11 3LW, UK
UKHW022314250926
475702UK00046B/782
* 9 7 8 1 0 3 2 3 8 5 7 9 2 *